Guide to the LEED® AP Interior Design and Construction (ID+C) Exam

GUIDE TO THE LEED® AP

Interior Design and Construction (ID+C) Exam

Michelle Cottrell, LEED AP ID+C

John Wiley & Sons, Inc.

This book is printed on acid-free paper. ∞

Copyright © 2012 by John Wiley & Sons, Inc. All rights reserved.

Published by John Wiley & Sons, Inc., Hoboken, New Jersey.

Published simultaneously in Canada.

No part of this publication may be reproduced, stored in a retrieval system, or transmitted in any form or by any means, electronic, mechanical, photocopying, recording, scanning, or otherwise, except as permitted under Section 107 or 108 of the 1976 United States Copyright Act, without either the prior written permission of the Publisher, or authorization through payment of the appropriate per-copy fee to the Copyright Clearance Center, Inc., 222 Rosewood Drive, Danvers, MA 01923, 978-750-8400, fax 978-646-8600, or on the web at www.copyright.com. Requests to the Publisher for permission should be addressed to the Permissions Department, John Wiley & Sons, Inc., 111 River Street, Hoboken, NJ 07030, 201-748-6011, fax 201-748-6008, or online at www.wiley.com/go/permissions.

Limit of Liability/Disclaimer of Warranty: While the publisher and author have used their best efforts in preparing this book, they make no representations or warranties with respect to the accuracy or completeness of the contents of this book and specifically disclaim any implied warranties of merchantability or fitness for a particular purpose. No warranty may be created or extended by sales representatives or written sales materials. The advice and strategies contained herein may not be suitable for your situation. You should consult with a professional where appropriate. Neither the publisher nor author shall be liable for any loss of profit or any other commercial damages, including but not limited to special, incidental, consequential, or other damages.

For general information on our other products and services, or technical support, please contact our Customer Care Department within the United States at 800-762-2974, outside the United States at 317-572-3993 or fax 317-572-4002.

Wiley publishes in a variety of print and electronic formats and by print-on-demand. Some material included with standard print versions of this book may not be included in e-books or in print-on-demand. If this book refers to media such as a CD or DVD that is not included in the version you purchased, you may download this material at http:/booksupport.wiley.com. For more information about Wiley products, visit www.wiley.com.

For more information about Wiley products, visit our Web site at www.wiley.com.

Library of Congress Cataloging-in-Publication Data:

Cottrell, Michelle.

Guide to the LEED AP interior design and construction (ID+C) exam / Michelle Cottrell.   1.

    p. cm. — (Wiley series in sustainable design; 34)

Includes index.

ISBN 978-1-118-01749-4 (pbk.); ISBN 978-1-118-16465-5 (ebk); ISBN 978-1-118-16466-2 (ebk); ISBN 978-1-118-16668-0 (ebk); ISBN 978-1-118-16669-7 (ebk); ISBN 978-1-118-16670-3 (ebk)

1. Leadership in Energy and Environmental Design Green Building Rating System–Examinations–Study guides. 2. Sustainable construction–Examinations–Study guides. 3. Sustainable buildings–Examinations–Study guides. I. Title.

TH880.C6785 2012 720'.47076–dc23

2011028936

Printed in the United States of America

10 9 8 7 6 5 4 3 2 1

Contents

Acknowledgments ix

Introduction xi

PART I: RAMPING UP   1

CHAPTER 1 UNDERSTANDING THE CREDENTIALING PROCESS _____ 3

The Tiers of the Credentialing Process   3

   The First Tier of the Credentialing System: LEED Green Associate   3

   The Second Tier of the Credentialing System: LEED Accredited Professional with Specialty   4

   The Third Tier of the Credentialing System: LEED Fellow   4

The Application Process   5

Apply!   6

Register!   6

Schedule!   7

Why Earn LEED Credentials?   7

CHAPTER 2 SUSTAINABILITY AND LEED BASICS REVIEW _____ 9

The Benefits of Green Buildings   10

   The Triple Bottom Line   11

The Design and Construction Process   13

   The Project Team Members   13

   Conventional Projects versus the Integrative Design Approach   13

Do Green Buildings Cost More?   16

USGBC and GBCI   17

CHAPTER 3 THE LEED FOR COMMERCIAL INTERIORS RATING SYSTEM _____ 21

LEED for Commercial Interiors (LEED CI)   21
The Categories of LEED   22
Prerequisites and Credits   22
    Credit Weightings   24
The LEED Certification Process   26
    LEED-Online   26
    Registration   27
    Credit Form   29
    Credit Interpretation Requests and Rulings   30
    Certification Review   30

PART II: DIVING IN: THE STRATEGIES AND TECHNOLOGIES OF LEED CI   37

CHAPTER 4 SUSTAINABLE SITES _____ 39

Site Selection   40
    Site Selection in Relation to LEED Compliance   41
Transportation   60
    Transportation in Relation to LEED Compliance   60
SS Credit Worksheets   68

CHAPTER 5 WATER EFFICIENCY _____ 101

Indoor Water Use   102
    Indoor Water Use in Relation to LEED Compliance   102
WE Study Worksheets   107

CHAPTER 6 ENERGY AND ATMOSPHERE _____ 109

Energy Efficiency   110
    Energy Efficiency in Relation to LEED Compliance   115
Tracking Energy Consumption   120
    Tracking Energy Consumption in Relation to LEED Compliance   120
Managing Refrigerants   124
    Managing Refrigerants in Relation to LEED Compliance   124
Renewable Energy   125
EA Study Worksheets   129

CHAPTER 7 MATERIALS AND RESOURCES — 147

Salvaged Materials and Material Reuse   149

Salvaged Materials and Material Reuse in Relation to LEED Compliance   150

Material Selection   152

Material Selection in Relation to LEED Compliance   154

Waste Management   159

Waste Management in Relation to LEED Compliance   160

MR Study Worksheets   164

CHAPTER 8 INDOOR ENVIRONMENTAL QUALITY — 185

Indoor Air Quality   187

Ventilation Strategies in Relation to LEED Compliance   187

IAQ Practices during Construction in Relation to LEED Compliance   190

Prevention and Segregation Methods in Relation to LEED Compliance   197

Thermal Comfort   199

Thermal Comfort in Relation to LEED Compliance   199

Lighting   202

Lighting in Relation to LEED Compliance   204

EQ Study Worksheets   209

CHAPTER 9 INNOVATION IN DESIGN AND REGIONAL PRIORITY — 243

Innovation in Design   243

ID Credit 1: Innovation or Exemplary Performance   243

ID Credit 2: LEED Accredited Professional   245

Regional Priority   245

ID Study Worksheets   250

PART III: STUDY TIPS AND APPENDICES   255

CHAPTER 10 STUDY TIPS   257

Preparing for the LEED AP ID+C Exam: Week Eight   257

Practice Exam Approach   258

The Testing Center Environment   258

Exam Structure 259
    When at the Testing Center 260
    Exam Scoring 260
    After the Exam 261

APPENDICES 263

    A. LEED CI Rating System 263
    B. Minimum Program Requirements (MPRs) for LEED CI 264
    C. LEED Certification Process 266
    D. Main Category Summaries 268
    E. Related Prerequisites and Credits 270
    F. Sample LEED CI Scorecard 272
    G. Exemplary Performance Opportunities – LEED CI v3.0 273
    H. Referenced Standards – LEED CI v3.0 276
    I. Answers to Quiz Question 280
    J. Abbreviations and Acronyms 290
    K. Sample Credit 294

CREDITS 295

INDEX 297

SAMPLE FLASHCARDS

Acknowledgments

To all my students, thank you so much for all the insight for the content for this study guide. Your questions and eagerness to learn continues to inspire me.

I would like to thank each of the image contributors, as the exam prep series would not be the same without your added visual integrity. Each of you helped to maintain my excitement about the book, as the readers will appreciate as well. Preparing for the exam with the added help of your images will greatly help them to remember the strategies of the rating system.

Zach Rose, assoc. AIA, LEED AP, LEED Green Associate, thank you for your support for not only this book and the others but for your continued encouragement and understanding. Thank you to the rest of my team at Green Education Services for your excitement and interest in the exam prep series!

Kathryn Malm Bourgoine, senior acquisitions editor at John Wiley & Sons, I enjoyed the opportunity to work with you and look forward to continuing my writing efforts with Wiley. Kerstin, Amy, Nancy, and Mike—your support during production has always made the process streamlined and enjoyable. Penny and Justin—thanks for your marketing wisdom! Thank you Lisa Ryan at Stellar Searches for helping me with yet another index!

A tremendous thank you to my family and friends for all of your support, motivation, interest, and patience! I really cannot thank you all enough!

Mom, thank you for taking care of Izzy while I was on the road and for always sending me home with your delicious meals!

Introduction

Guide to the LEED® AP Interior Design and Construction (ID+C) Exam is the resource to prepare for the Leadership in Energy and Environmental Design (LEED®) Accredited Professional (AP) Interior Design + Construction (ID+C) exam. This exam prep guide provides a road map to studying for the LEED AP ID+C exam as administered by Green Building Certification Institute (GBCI). The *Guide to the LEED AP Interior Design and Construction (ID+C) Exam* is aimed at those professionals seeking more information about the basic knowledge and understanding that is required in order to pass the exam and earn the LEED AP ID+C accreditation.

As a means to introduce myself, I am a LEED consultant and an education provider, focused on sustainable design and building operation concepts. I have traveled the country helping hundreds of students to prepare for the LEED Green Associate and LEED AP exams. The LEED AP ID+C classes typically are one-day seminars that review all of the information as presented in this book. During these classes, I share my LEED project experiences and study tips in order to help make sense of this challenging information and present it in a logical format to help streamline the studying efforts for my students. This book breaks down the difficult information to be retained into a coherent and straightforward approach, as compared to simply repeating what would be found in the study reference material outlined by GBCI.

> **TIP** Keep an eye out for these **STUDY TIPS**! as they will point out the intricacies and nuances to remember.

EXAM PREP GUIDE STRUCTURE

Guide to the LEED AP Interior Design and Construction (ID+C) Exam is organized into three parts as a method to break down the information one can expect to see on the exam. First, an introduction is needed to review the concepts and process covered in the LEED Green Associate exam in order to then understand the next part, which covers the technologies and strategies to implement as detailed in each prerequisite and credit. Finally, the appendices include charts and diagrams summarizing the critical information, as well as other resources to narrow down the amount of information to be studied as preparation to sit for and pass the LEED AP ID+C exam. The composition of the book is as follows:

Part I: Ramping Up *is composed of the following information:*

- Chapter 1: Understanding the Credentialing Process
- Chapter 2: Sustainability and LEED Basics Review
- Chapter 3: The LEED for Commercial Interiors Rating System

TIP Be sure to review the eligibility requirements described in Chapter 1 to apply for the LEED AP ID+C exam.

Part II: Diving In: The Strategies and Technologies of LEED CI details the overlying concepts of the primary categories and the strategies to achieve each basic concept within the rating system. Part II is focused on organizing the information to remember and also to provide the concepts behind each prerequisite and credit. It is intended to work in tandem with the coordinating study worksheets included at the end of each chapter to help you remember the details, such as the intents, requirements, documentation requirements, and required calculations of each prerequisite and credit. In Part II the following LEED categories are reviewed:

- Chapter 4: Sustainable Sites
- Chapter 5: Water Efficiency
- Chapter 6: Energy and Atmosphere
- Chapter 7: Materials and Resources
- Chapter 8: Indoor Environmental Quality
- Chapter 9: Innovation in Design and Regional Priority

Part III: Study Tips and Appendices is dedicated to summarizing the critical information, details, and concepts to retain, as well as providing an overview of the testing center environment. The appendices include additional resources to help summarize the information presented in Parts I and II, such as scorecards and summary charts.

STUDY TIPS! are located throughout the book as tools to help stay focused on the pertinent information. They will include things to remember and point out side note type of information. Sample exam questions (in terms of format and content) are also found in this book, as well as more basic quiz questions placed sporadically throughout.

Be sure to spot these **FLASHCARD TIPS!** to create flashcards along the way. Use the white cards for Part I and the color-coded ones for Part II.

While reading through this book, be sure to also keep an eye out for **FLASHCARD TIPS!**, as they will help to distinguish the important aspects for the exam and act as an indicator to create critical flashcards. All of the FLASHCARD TIPS! referenced throughout the book are collected at the end, following the index, although it is suggested that you create your own to enhance your studying. It is recommended that you purchase plain white note cards, as well as the color-coded note cards (i.e., pink, yellow, blue, green, and purple). Use the white ones for the information to be covered in Part I and the color-coded cards for Part II of this exam prep book. The FLASHCARD TIPS! suggest a starting point for flashcard creation, but feel free to make more as needed. If you decide to make your own with the help of the FLASHCARD TIPS!, be sure to refer to flashcards at the end for some additional flashcard suggestions. If you decide to use the flashcards from the book and not make your own, you can always use markers or highlighters to color-code them for streamlined studying.

Be sure to look out for these **BAIT TIPS!** as well. These tips will reinforce the important concepts and **B**ring **A**ll of **I**t **T**ogether as synergies and trade-offs are pointed out for green building strategies and technologies.

One of the main concepts of sustainable design is the integrative fashion in which green buildings are designed and constructed. It is critical to understand how strategies and technologies have synergies and trade-offs. For example, green roofs can have an impact on a construction budget but can help save on operational energy costs, which may present a breakeven or surplus. Green roofs, as seen in Figure I.1, also have synergistic qualities because they can not only help reduce the heat burden on a building, but also help to manage stormwater. These types of concepts will be discussed in greater detail in Part II of this exam prep guide, but for now be sure to look for these **BAIT TIPS!** throughout Part II to help bring the concepts together.

Figure I.1 Holy Wisdom Monastery project in Madison, Wisconsin, by Hoffman, LLC, reduces the amount of stormwater runoff from the site, increases the amount of open space, promotes biodiversity, reduces cooling loads, and reduces the impacts of the urban heat-island effect by implementing a vegetated roof.
Photo courtesy of Hoffman, LLC

STUDY SCHEDULE

Week	Chapters	Pages
1	Part I: Ramping Up (Chapters 1–3)	1-38
2	Part II: Sustainable Sites (Chapter 4)	39-100
3	Part II: Water Efficiency (Chapter 5)	101-108
4	Part II: Energy and Atmosphere (Chapter 6)	109-146
5	Part II: Materials and Resources (Chapter 7)	147-184
6	Part II: Indoor Environmental Quality (Chapter 8)	185-242
7	Part II: Innovation in Design and Regional Priority (Chapter 9) and Part III: Study Tips	243-262
8	Study flashcards, rewrite your cheat sheet a few times, and take online practice exams	
9	Register and take LEED AP ID+C Exam!!	

As the preceding table shows, it is recommended that you read through Parts I and II of this exam prep book within seven weeks. Introductory terminology from Part I should be absorbed to get on the right path to understand the more critical exam-oriented information presented in Part II. The goal is to create a complete set of flashcards during the first seven weeks while reading through the material, thus allowing the following week (eighth week of studying) to focus on memorizing and studying the flashcards, followed by taking a few online practice exams, which are available at www.GreenEDU.com.

Although the exam format and structure will be reviewed in Part III of this book, there is one component that should be revealed up front. When you are at the testing center and about to take the exam, there will be an opportunity

 TIP After taking some practice exams, you may want to add to your cheat sheet and/or your flashcards.

to make a "cheat sheet" of sorts. Although you will not be allowed to bring any paper, books, or pencils into the exam area, you will be supplied with blank paper and a pencil (or a dry-erase board and a marker). So now that you know this opportunity is there, let's take advantage of it! Therefore, as a concept, strategy, referenced standard, or requirement is presented in this exam prep guide, make note of it on one single sheet of paper. At the end of Part II, this "cheat sheet" should be reviewed and then rewritten with the critical information you determine that you might forget during the exam. You are the only one who knows your weaknesses in terms of the information you need to learn—I can only make recommendations and suggestions. During Week Seven, you should rewrite your cheat sheet two to three more times. The more you write and rewrite your cheat sheet, the better chance you will have for actually retaining the information. It is also advised that you monitor the time it takes to generate your cheat sheet, as time will be limited on exam day.

If you maintain the recommended study schedule, eight weeks from now a set of flashcards will be created and your cheat sheet started. You will then have one week of consistent studying time to focus on the material in your flashcards. After studying your flashcards, it is recommended that you take a few online practice exams to test your knowledge. The approach to these sample exams is described in Part III, Chapter 10, of this book, including the next steps for the cheat sheet. After a few practice exams, an assessment of your preparation should be completed to determine if you are ready for the exam. Your exam date should be scheduled at that time.

Before focusing on the exam material, be sure to read through Chapter 1 to understand the application requirements of the LEED AP ID+C exam to ensure your eligibility and to understand the exam application process.

Guide to the
LEED® AP Interior Design and Construction (ID+C) Exam

PART I

RAMPING UP

CHAPTER 1
UNDERSTANDING THE CREDENTIALING PROCESS

BEFORE DIVING INTO THE EFFORT OF STUDYING and preparing for the Leadership in Energy and Environmental Design (LEED®) Accredited Professional (AP) Interior Design and Construction (ID+C) exam, there are quite a few things to review to ensure your eligibility. Whenever I teach an exam prep course, this topic is not typically addressed until the end of the class, as it is easier to digest at that point; but it is important to present this information here in the first chapter, to make sure the test is applicable to and appropriate for you. This chapter will provide the important concepts of the tiered credentialing system to ensure that the components, the exam application process, and the requirements for eligibility are understood.

This initial information begins with the new credentialing system for LEED accreditation, as it involves three tiers:

1. LEED Green Associate
2. LEED Accredited Professional (AP) with Specialty
3. LEED Fellow

THE TIERS OF THE CREDENTIALING PROCESS

The first step of comprehending the credentialing process begins with a brief understanding of the basics of LEED. LEED is the acronym for Leadership in Energy and Environmental Design, signifying a green building rating system designed to evaluate projects and award them certification based on performance. The U.S. Green Building Council's (USGBC®) website indicates that LEED has become the "nationally accepted benchmark for the design, construction, and operation of high performance green buildings."[1] USGBC created the first LEED Green Building Rating System back in the 1990s as a tool for the public and private commercial real estate markets to help evaluate the performance of the built environment.

 TIP Notice the LEED acronym does **not** contain an "S" at the end. Therefore, please note this first lesson: when referring to LEED, please do not say "LEEDS," as it is quite important to refer to the acronym correctly.

The First Tier of the Credentialing System: LEED Green Associate

The **LEED Green Associate** tier is applicable for professionals with a basic understanding of green building systems and technologies. These professionals have been tested on the key components of the LEED rating systems and the certification process. This level of credentialing is the first step to becoming a LEED AP.

The Green Associate exam is geared toward all professionals involved in the world of sustainable design, construction, and operations beyond just the typical architecture and engineering design professionals. Therefore, the exam is available for lawyers, accountants, contractors, manufacturers, owners, and developers, as well. Any professional who works in the field of sustainable design and green building is eligible to sit for the exam, especially those with LEED project experience. For those who wish to sit without LEED project experience or who are not employed in a sustainable field of work, participating in an educational course focused on sustainable design would qualify instead.

The Second Tier of the Credentialing System: LEED Accredited Professional with Specialty

The next tier, **LEED AP with Specialty**, is divided into five types (of specialties):

1. *LEED AP Building Design + Construction* (BD+C). This exam includes concepts related to new construction and major renovations, core and shell projects, and schools. This specialty will also cover retail and health care applications in the future.
2. *LEED AP Interior Design + Construction* (ID+C). This exam contains questions related to tenant improvement and fit-out project knowledge for commercial interior and retail professionals.
3. *LEED AP Operations + Maintenance* (O+M). This exam covers existing building project knowledge specific to operations and maintenance issues.
4. *LEED AP Homes.* This exam applies to professionals practicing in the residential market.
5. *LEED AP Neighborhood Development* (ND). This exam tests whole or partial neighborhood development project knowledge.

Because the LEED Green Associate credentialing tier is the first step to obtaining LEED AP status, the LEED AP exams are thought of in a two-part exam process beginning with the LEED Green Associate exam. You have the option to decide whether you wish to take both exams in one day or break the exam into two different testing appointments. The exams are quite challenging and mind intensive, and can be exhausting, so bear this in mind when deciding on which option to pursue.

LEED project experience is required in order to be able to sit for any of the LEED AP specialty exams. These exams cover more in-depth knowledge of each of the prerequisites and credits; the requirements to comply, including documentation and calculations; and the technologies involved with the corresponding rating system. These exams are, therefore, applicable for those professionals working on LEED registered projects or those who have worked on project within the last three years that earned certification.

The Third Tier of the Credentialing System: LEED Fellow

Finally, the third tier of the credentialing system, **LEED Fellow**, is the highest level of credentialing. It is meant to signify a demonstration of accomplishments, experience, and proficiency within the sustainable design and construction community. These individuals will have contributed to the continued development of the green building industry for at least eight years as an AP and will need to be nominated for the credential.

THE APPLICATION PROCESS

Now that there is an understanding about the three tiers of the credentialing system, whom each tier is geared for, and the eligibility requirements of each exam type, it is time to review the process for applying for the exam. The first step involves visiting the Green Building Certification Institute (GBCI®) website at www.gbci.org and downloading the *LEED AP ID+C Candidate Handbook* found in the Professional Credentials section of the website.

Each of the candidate handbooks details the following information:

- Study materials, including the exam format, timing, references, and sample questions
- How to apply for the exam, including the application period, eligibility requirements, and exam fees
- How to schedule your exam once your eligibility is confirmed, including confirmation, and canceling, and rescheduling your test date
- A pre-exam checklist
- What to expect on the day of your exam, including name requirements, scoring, and testing center regulations
- What to do after your exam, including the Credentialing Maintenance Program (i.e., continuing education requirements) and certificates

TIP GBCI updates the candidate handbooks for each of the exam types at the beginning of each month, so make sure to have the most current version.

Although the intention of this exam prep book is to consolidate all of the information needed to prepare for the LEED AP ID+C exam, some of the references are updated from time to time. Therefore, this book contains information similar to that found in the handbooks, for added efficiency, but you are best advised to refer to the latest version of the handbook appropriate to the LEED AP ID+C credential for the most up-to-date exam information, especially as related to the exam application process.

In order to understand why a different organization (other than USGBC) is the resource for information for LEED professional credentials and is the destination website to apply for the exam, the role of GBCI is presented next. As previously mentioned, LEED is an independent, third-party verified, voluntary rating system, and in order to be in compliance with ANSI/ISO/IEC 17024's accreditation requirements, USGBC created GBCI to separate the rating system development from the credentialing program. GBCI is now responsible for LEED project certification and professional credentialing, while USGBC is responsible for the development of the LEED rating systems and educating the industry for continuing efforts to help evolve the green building movement and, thus, transform the market. Case in point: the USGBC website should be visited to obtain information about each rating system or to purchase reference guides, whereas the GBCI website should be the resource for information about taking an exam or to register a project seeking certification and learn more about the certification process. Chapter 2 defines the roles of the two organizations in more detail.

TIP It is important to refer not only to LEED correctly but also to projects and professionals. Remember, buildings are **certified** and people are **accredited**. People will never be able to become LEED certified professionals—remember, there are LEED APs and not LEED CPs.

Additionally, LEED *certification* is meant for projects and buildings, not products. Not only will a LEED certified professional not be found but also neither will a LEED certified chair, air-conditioning unit, appliance, paint, or glue.

Figure 1.1 The steps to register for the LEED AP ID+C exam.

 TIP It is *critical* to sign up and create the account with GBCI consistent with the account holder's name as it appears on the identification to be used to check in at the testing center. If they do not match on the day of the exam, exam fees may be lost, as the opportunity to take the exam may be forfeited. If your existing account with USGBC is not consistent with your identification, refer to the *LEED AP ID+C Handbook* for instructions on how to update your account information.

Once the handbook is downloaded and reviewed, the next step includes establishing an account with the GBCI website. If an account already exists with USGBC, the same one can be used for GBCI's website, as they are the same. Therefore, once an account is established with GBCI, the same login information will work on USGBC's website. Should a new account need to be established, navigate to the "Log In to My Credentials" section of the GBCI website to create an account.

APPLY!

Once an account is established with GBCI, the next step is to apply for eligibility. On the GBCI website, visit the "My Credentials" section to begin the process after logging in. Make sure the profile is correct and select the intended credentialing path. If you have already passed the LEED Green Associate exam, you will only need to have worked on a LEED project within the past three years to be eligible to sit for the LEED AP ID+C exam. Be sure to refer to the candidate handbook for more information about the requirements for project participation. Next, enter in the project information, such as name, location, and rating system in which the project was registered or certified under and upload the documentation proving eligibility as described in the *LEED AP ID+C Candidate Handbook*. You will then need to pay the nonrefundable $100 application fee. Within seven days, you should receive an email indicating approval or denial, in order to move to the next step of the exam registration process. "Five to seven percent of all applications will be audited; you will be notified immediately if you are chosen for an audit and will be notified of your eligibility within seven days."[2] Should you receive indication of ineligibility, you are required to wait 90 days to apply again.

REGISTER!

Your application is valid for up to one year, once approval notification is received. At this point, the next step of registering for the exam should be seen as an option within the "My Credentials" section of the website. Here, verification is required for the test to be registered for and confirmation of membership status. Remember, USGBC national company members can take advantage of reduced exam fees. This means the company in which you work for must be a national member of USGBC. To receive the discount, you will need to ensure your USGBC/GBCI account profile contains the Corporate ID associated with the company in which you for.

SCHEDULE!

The next step is scheduling an appointment to take the exam at a Prometric testing center. As stated previously, it is advised that you defer selecting an exam date until you are further along in the preparation for the exam. In the introduction of this exam prep book, a study and reading schedule is suggested. It is highly recommended that you start studying and determine your level of knowledge of the test content before scheduling an exam date.

When you are ready to schedule an exam date, visit www.prometric.com/gbci, or if at the GBCI website, follow the links to the Prometric website to schedule a day to take the exam, from the "My Credentials" section. Remember, the eligibility code from GBCI is required to schedule an exam date. After an exam date is scheduled, a confirmation code is displayed on the screen. Keep this code! This code will be needed should the selected exam date need to be canceled, confirmed, or rescheduled with Prometric. A confirmation email containing your confirmation code will be sent from Prometric shortly after scheduling.

In addition, it is important to remember that candidates will have three allowed testing attempts per one-year application period. In the event that a retake is necessary (even though this is not the plan!), test takers will need only to pay an additional fee for the exam and not the application fee. Refer to the *LEED AP ID+C Handbook* for more information on this rule.

TIP To reschedule or cancel an exam date, please consult the *LEED AP ID+C Candidate Handbook* for explicit instructions. GBCI is quite meticulous about the procedure, so it is advised that you be aware of the details to avoid risking a loss in fees paid.

WHY EARN LEED CREDENTIALS?

Green buildings are evaluated according to the triple bottom criteria (social, economic, and environmental); deciding whether to earn LEED credentials can be approached in the same fashion, as there are individual, employer, and industry benefits to examine. From an individual standpoint, earning the LEED AP ID+C credential will grant professionals a differentiator to market themselves to a potential employer or clients, provide them with exposure on the GBCI website database of LEED professionals, and earn them a certificate to display and recognition as professionals in the LEED certification process. An employer would also benefit by earning the eligibility to participate in LEED projects, as more projects are requiring LEED credentials for team members; building the firm's credentials when responding to requests for proposals (RFPs) and requests for qualifications (RFQs); and having the opportunity to encourage other staff members to aim for the same credential to help the firm to evolve. Finally, the market would also benefit as more professionals earn the LEED AP ID+C credential by helping the built environment to become more sustainable and the market to evolve, transform, and grow.

CHAPTER 2
SUSTAINABILITY AND LEED BASICS REVIEW

AS MENTIONED EARLIER, IT IS CRITICAL TO BE ON THE RIGHT PATH by remembering the basic concepts tested on the Leadership in Energy and Environmental Design (LEED®) Green Associate exam before jumping into the details of the LEED categories as seen on the LEED AP ID+C exam. Therefore, sustainability and green building are described and detailed as a starting point. What is sustainability? The Wikipedia website refers to the concept as the "the capacity to endure."[1] For the purposes of LEED, it is important to take a step further beyond sustainability and think of **sustainable design** and development. Although the definition is not universally accepted, the Brundtland Commission of the United Nations' website is cited (for the purposes of the exam and LEED) for their definition: "development that meets the needs of the present without compromising the ability of future generations to meet their own needs."[2]

Within the design industry, sustainable design and sustainable building concepts are interchangeable with the term **green building**—the next vocabulary word to become familiar with. When referring to green buildings, it is understood that the buildings are sensitive to the environment, but one might wonder, how exactly? Green buildings are more efficient and use resources wisely, as they take energy, water, and materials into account (Figure 2.1). But one might ask, "How do they use resources more efficiently?" To answer this question, it is important to think of the different aspects of a building, for instance:

- *Site selection.* Is the project a redevelopment in an urban area, or does it support urban sprawl? How close is the project to public transportation to reduce the number of cars coming and going? How will the building need to be situated in order to take advantage of the natural breezes for ventilation and daylight to reduce the need for artificial lighting within the building?

- *Design of the building systems, such as mechanical equipment, the building envelope, and lighting systems.* How do they work together? Were they designed independently of each other? Is the heat emitted from the lighting fixtures accounted for? Are there gaps in the envelope that allow conditioned air to escape?

- *Construction processes.* Think about the people on site during construction—are they being exposed to harmful fumes and gases? Are precautions being taken to reduce the chances for mold growth or other contaminants?

- *Operations of the building.* What kind of items are purchased to support business? What about cleaning procedures?

- *Maintenance.* When was the last time equipment was tested to ensure that it is performing appropriately? Are there procedures in place to monitor for leaks?

Figure 2.1 Reasor's Supermarket in Owassa, Oklahoma, incorporates daylighting strategies and polished concrete floors, together helping the project to earn multiple LEED credits within different categories, including Materials and Resources, Energy and Atmosphere, and Indoor Environmental Quality. *Photo courtesy of L&M Construction Chemicals, Inc.*

TIP When thinking of green buildings, it is important to think of not only how the building is designed to function and how it is constructed but also the environmental impacts from operations and maintenance.

- *Waste management.* How is construction waste addressed? What about the garbage generated during operations? Is it going to the landfill? Who knows where those containers are going?!

THE BENEFITS OF GREEN BUILDINGS

Hopefully, the previous questions started to generate some thoughts of what is involved with green buildings. If not, maybe evaluating the benefits of green buildings might help; beginning with a review of the traditional buildings statistics and how they impact our planet. The U.S. Green Building Council (USGBC®) has compiled information from the Energy Information Administration and the U.S. Geological Survey on the impacts of buildings on our natural resources in the United States. The USGBC website reports the following statistics for *conventionally* designed and built buildings:

    39 percent primary energy use

    72 percent electricity consumption

38 percent carbon dioxide (CO_2) emissions

14 percent potable water consumption[3]

It is important to digest the 39 percent CO_2 emissions statistic, as this percentage puts buildings at the top of the list, followed by transportation and industry. Buildings have a bigger impact on greenhouse gas emissions—the biggest, actually! These statistics have pushed the market to find better ways to design, construct, and operate buildings.

 TIP Write it, read it, say it, and hear it as many times as possible. The more senses you involve in your studying efforts, the more you information you will be able to retain.

When looking at the statistics for green buildings, including LEED-certified buildings, the General Services Administration (GSA) indicates that these projects have been able to achieve the following:

26 percent energy use reduction

40 percent water use reduction

70 percent solid waste reduction

13 percent reduction in maintenance costs[4]

These percentages reflect the benefits in the economic bottom line, but these green buildings have also reduced their impact on the environment, as well as demonstrated an improved indoor environment (such as air quality) and contribution to the community. Indoor air quality is extremely important when analyzing the benefits of green buildings, as the Environmental Protection Agency (EPA) reports Americans "spend, on average, 90 percent or more of their time indoors."[5] Green buildings have resulted in 27 percent higher levels of satisfaction[6] and allowed students the opportunity to perform better.[7]

The Triple Bottom Line

The USGBC website summarizes the benefits of green buildings according to three components: environmental, economic, and health and community benefits, as shown in Table 2.1. In the green building industry, these three concepts are defined as the *triple bottom line* (Figure 2.2). A conventional project usually assesses only the singular component of the economic prosperity for the project.

Table 2.1 The Benefits of Green Buildings[8]

Environmental Benefits	Enhance and protect ecosystems and biodiversity
	Improve air and water quality
	Reduce solid waste
	Conserve natural resources
Economic Benefits	Reduce operating costs
	Enhance asset value and profits
	Improve employee productivity and satisfaction
	Optimize life-cycle economic performance
Health and Community Benefits	Improve air, thermal, and acoustic environments
	Enhance occupant comfort and health
	Minimize strain on local infrastructure
	Contribute to overall quality of life

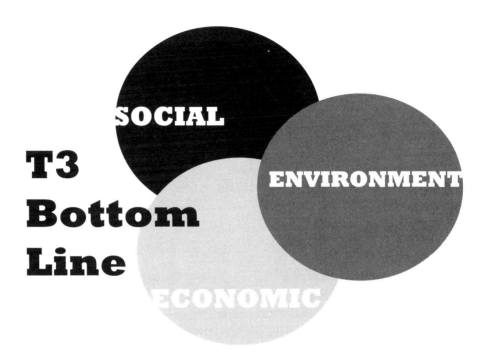

Figure 2.2 The triple bottom line.

However, when determining the goals for a project seeking LEED certification, the process typically begins with assessing the goals in comparison to the *triple bottom line* values. For example, should a client wish to install a green roof on their building, the team would assess the financial implications as compared to the environmental impacts versus the community benefits. These types of details will be discussed later, but understanding the three types of benefits is important at this time.

> **TIP** These questions are formatted just as they would be on the exam. Notice that the question indicates how many answers to select. The proper number of correct answers is required on the exam, as partial credit is **not** awarded.

QUIZ TIME!

Q2.1. Which of the following is an environmental benefit of green building? (Choose one)

    A. Conserve natural resources

    B. Reduce solid waste

    C. Improve air and water quality

    D. Enhance and protect ecosystems and biodiversity

    E. All of the above

> **TIP** The answers to all of the quiz questions can be found in Appendix I.

Q2.2. How much time, on average, do Americans spend indoors? (Choose one)

    A. 10 percent

    B. 90 percent

    C. 65 percent

    D. 35 percent

Q2.3. According to the Department of Energy's website, space heating is the largest energy use in the United States, followed by lighting. True or False?

A. True

B. False

Q2.4. Which of the following describes a high-performance green building? (Choose one)

A. Conserves water and energy

B. Use spaces, materials, and resources efficiently

C. Minimizes construction waste

D. Creates a healthy indoor environment

E. All of the above

> **TIP** Don't worry if some of the questions presented are unfamiliar territory. The questions in this book are meant to present you with new information to learn from and to prepare you for the real exam, as there is bound to be information presented for which you will need to use the process of elimination to determine the best answers.

THE DESIGN AND CONSTRUCTION PROCESS

The Project Team Members

Understanding the processes of design and construction, from a traditional or conventional standpoint as opposed to that of sustainable projects, begins with an understanding of the players involved in the process:

Architect. Responsible for the design of green building strategies, including overall site planning and interior spaces

MEP engineer. Responsible for the design of the energy and water systems of a building, more specifically, the mechanical, electrical, and plumbing components, including thermal impacts

Landscape architect. Responsible for the selection of trees and plants, the impacts of shading, and water efficiency for irrigation; also responsible for vegetated roof design

Civil engineer. Responsible for site design, including stormwater management, open space requirements, and site protection

Contractor. Typically referred to as the GC, short for general contractor; responsible for the demolition (if required) and construction of a facility, including site work

Facility manager. Also referred to as a building engineer; responsible for maintaining a building and its site during operations

Commissioning authority (CxA). Responsible for the commissioning process, including a drawing review during design and equipment installation and a performance review during construction

Owner. Defines the triple bottom line goals and selects the team members for a project; can be a developer and does not have to be the end user

End users/occupants. The inhabitants of a building and, therefore, should be the main priority when designing for comfort and productivity

 For those not familiar with the professionals involved, create flashcards for each to remember their roles and importance.

Conventional Projects versus the Integrative Design Approach

The next step in understanding the process of design and construction involves comprehending the different types of projects, as well as the difference in the approach for conventionally designed buildings versus the integrative approach pursued by sustainably designed projects. Projects pursuing LEED certification are approached differently from conventional projects, as they use an integrative process that begins at the onset of the project, or as early as possible during design.

Although there are substantial differences between conventional and sustainably driven projects, they both consist of the same phases for the design and construction processes.

Phases of the Traditional Project Delivery

- Predesign/Programming
- Schematic Design phase
- Design Development phase
- Construction Documents phase
- Agency Permit/Bidding
- Construction
- Substantial Completion
- Final Completion
- Certificate of Occupancy

Phases of the Integrated Project Delivery

- Conceptualization
- Criteria Design
- Detailed Design
- Implementation Documents
- Agency Coordination/Final Buyout
- Construction
- Substantial Completion
- Final Completion
- Certificate of Occupancy

> **TIP** Notice when all the players are introduced to the project and how they all work in a linear and independent fashion for a conventionally designed project.

The key difference of the phases depends on who is involved and when, when comparing a traditionally designed and constructed building to one that is designed with sustainable initiatives. For example, with a traditionally designed project, an owner may hire a civil engineering or environmental team once they select a piece of property. Once the environmental reports are completed and they have an idea of how their building can fit on the site, the site plan is handed off to an architect. The architect then works with the owner to detail the program requirements (known as the **Programming** phase) and then begins to design the building (known as the **Schematic Design** phase). The architect then works with an engineering team (typically composed of mechanical, electrical, and plumbing engineers and a structural engineer, if needed, depending on the project type). These professionals typically work independently of each other to complete their tasks (known as the **Design Development** phase). Remember, with a traditionally designed project, the architect has already designed the building and is now handing off the plans to the engineers to fit the building systems into the building that was designed without their input. Once the basic design elements are established, each professional works to complete a set of **construction documents** (CDs). Notice that the responsibilities are segmented just as the communication is fragmented.

What happens next with the CDs varies with different project types. Typically, these documents are first issued for permit review by the local municipality. It is quite common for most projects to send the CDs out for bid to a number of contractors about the same time as the drawings are issued for permit review (known as a Design-Bid-Build project type), while other projects have the contractor engaged as one entity with the architect from the beginning (known as a Design-Build project type).

At this point in a Design-Bid-Build project type, the contractor is given a short period of time in which to evaluate the drawings and provide the owner with a fee to for demolition services (if required) and to construct the building, including site development work. The contractors are given an opportunity to submit questions (known as requests for information, or RFIs) about the requirements or design elements during this bidding process, but then they are held to the quote they provide. Remember, the contractor was not engaged during the previous design phases, so he or she is not familiar with the project and have to dive in quickly, sometimes making assumptions about the construction requirements. Most of the time, projects are awarded based on the lowest bid, but think about the implications of doing so. If the lowest bidder wins the job, where are they cutting corners? Is quality being compromised? Was a critical element omitted? No one likes to lose money, as that is just bad business, but is this really the best way to select a contractor?

Besides Design-Bid-Build and Design-Build, other project delivery types are Multiprime and Construction Manager at Risk.

Once the permit is received, the contractor is selected and the construction cost is agreed upon, the phases of the design process are over and the construction process begins. Just as the design process has four phases, the construction process does as well. **Construction** commences the process, traditionally with little involvement from the design team. The next phase, **Substantial Completion,** includes the final inspection process and when the owner issues a "punch list." The owner compiles a punch list while walking the space with the contractor and notes any problems requiring the contractor's attention. **Final Completion** is next, followed by the **Certification of Occupancy.** Once the Certificate of Occupancy is received, the building is then permitted to be occupied.

Remember, the integrative process needs to be implemented as early as possible in order for the true benefits to be achieved!

When compared to the traditional project delivery method, the integrative design process for sustainable design projects involves different phases of design and construction, as shown in the previous list, and remember the main differentiator is determined by the team members, particularly how and when they are involved. For a project seeking LEED certification, the owner may engage a number of consultants early in the process to assist in selecting the property or tenant space. He or she may retain an architect to evaluate the site for building orientation options to capitalize on natural ventilation or daylighting opportunities. He or she may hire a civil engineer to research the stormwater codes and to determine access to public transportation. A LEED consultant may be engaged to assist with evaluating the triple bottom line goals particular to a project site or tenant space. Think about the benefits of bringing the landscape architect and the civil engineer on board simultaneously so that they can work together to reveal the opportunities to use stormwater collection for irrigation needs. If the site were already determined, the owner would bring all of the consultants (including the general contractor) together to review the economic, social, and environmental

TIP Remember, an integrated project delivery (IPD) differs from a conventional project in terms of teams, process, risk, communications, agreement types, and phases.

goals collaboratively. This goal-setting meeting, or *charette,* is a key component of the first step of a sustainable project and is, therefore, part of the first design phase of Programming or Predesign, as the integrative process should be started as early as possible. This early start concept is graphically represented in the Macleamy Curve on page 21 of the *Integrated Project Delivery: A Guide.*

Another key difference with a green building project is the use of energy modeling and Building Information Modeling (BIM). These tools allow the design team to find efficiencies and conflicts with their design intentions. They can model the proposed building systems to evaluate and predict the performance of the components specific to the elements and the project's location and site. These technologies allow the design team to specify systems and equipment sized appropriately for the specific building. Because the tools allow for the project to be evaluated from a three-dimensional perspective, design teams will also have the opportunity to find conflicts with building components and systems. The design teams can even use these tools to determine the estimated energy and water savings as compared to implementing traditional building systems. These tools are used throughout the design phases to bring more efficiency to the project for all team members.

Projects utilizing an integrative design approach bring the entire team together early in the design process, thus allowing the opportunity for everyone to work more collectively, which can actually save time and money. A project's schedule can be reduced because the project's goals are reinforced throughout every step of the process. An integrated project delivery (IPD) avoids the "value engineering" aspect that can happen on a conventionally designed project. Value engineering (VE) can take place when the bidding contractors respond with a construction cost much higher than anticipated by the owner and design professionals. In response to this high price, the design team begins to remove design elements from the original scope of work to try to get the construction cost better aligned with the project budget. IPD projects avoid this inefficiency because the contractor is evaluating the elements and drawings continuously throughout each of the design phases.

TIP Remember, with an IPD the risks may be shared across the team, but so are the rewards!

In summary, traditionally designed projects differ from IPDs in terms of teams, process, risk, communications, agreement types, and phases. Remember, conventional project teams are fragmented, whereas green building teams work more collectively. An IPD project's process is more holistically approached, while a traditional project is more linear. The risk is separated with a fragmented, traditional project as compared to an IPD. In terms of communicating ideas and concepts, traditional projects are presented in a two-dimensional format, whereas sustainable projects work with BIM technologies to allow for the opportunity to find conflicts. Agreement types can vary, but with an IPD there is more collaboration to encourage a multilateral approach as compared to the unilateral approach of a conventional project. Finally, the phases change names from a traditional approach to an IPD.

DO GREEN BUILDINGS COST MORE?

TIP Chapter 7 discusses the environmental components of LCAs.

When assessing the cost of any type of project, it is important to understand the different types of costs involved. Traditionally, only two types of costs are detailed in a project's pro forma: hard costs and soft costs. Hard costs are defined as construction costs, including site work and demolition, whereas soft costs are related to the fees for professional services, including legal and design. Soft costs also include pre- and postconstruction-related expenses, such as insurance.

Green building projects take budgeting a step farther by including a **life-cycle assessment** (LCA) cost. LCAs include the purchase price, installation, operation, maintenance, and replacement costs for each technology and strategy proposed, to determine the appropriateness of the solution specific to the project.

USGBC has promoted many studies, including one from Davis Langdon (found in the references listed in the *LEED AP ID+C Candidate Handbook*), indicating that green building does not have to cost more. This is especially true if the project starts the process early in the design phases. It is also important to bridge the gap between capital and operating budgets to understand the value of green building technologies and strategies. For example, the first or up-front cost of installing photovoltaic panels, high-efficiency mechanical systems, or an indoor water wall to improve indoor air quality may not fit in a typical budget, but if the utility cost savings were considered and evaluated, either one might make more sense. Another case in point, first costs may also be higher in a traditionally designed project because of the lack of integration. For example, a mechanical engineer may specify a larger mechanical system than what is actually needed because he or she may not realize that high-performance windows were specified by the architect, along with building insulation with a higher R-value. Remember, the economic bottom line is important, but a green building project also evaluates the environmental and social impacts and benefits.

USGBC AND GBCI

Although Chapter 1 briefly introduced USGBC, it is important to remember the organization as "a 501(c)(3) nonprofit composed of leaders from every sector of the building industry working to promote buildings and communities that are environmentally responsible, profitable and healthy places to live and work," as posted on the USGBC website.[9] USGBC's mission statement is also listed on their website as "to transform the way buildings and communities are designed, built and operated, enabling an environmentally and socially responsible, healthy, and prosperous environment that improves the quality of life."[10] Remember also that USGBC created GBCI in January 2008 to "administer project certifications and professional credentials and certificates within the framework of the U.S. Green Building Council's Leadership in Energy and Environmental Design (LEED®) Green Building Rating Systems™," as indicated on the GBCI website.[11] GBCI's mission statement is stated as "to support a high level of competence in building methods for environmental efficiency through the development and administration of a formal program of certification and recertification."[12]

As indicated in Figure 2.3, USGBC is focused on developing the LEED Green Building Rating Systems, as well as providing education and research programs.

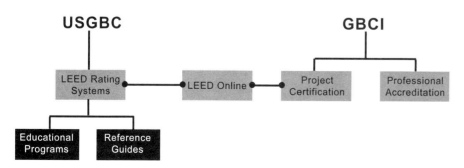

Figure 2.3 The roles of USGBC and GBCI.

Make a flashcard to remember what a TAG is.

In order to develop the rating systems, USGBC created a LEED Steering Committee composed of five technical advisory groups (TAGs) to help the main categories evolve. Eight regional councils are also a part of USGBC to help with the regional components of the rating systems.

GBCI was created in order to separate the rating system development from the certification and credentialing process. Therefore, GBCI is responsible for administering the process for projects seeking LEED certification and for professionals seeking accreditation credentials. GBCI administers the LEED certification process and are therefore responsible for managing the review process, determining a building's compliance with LEED standards, and establishing the level of certification for which they qualify.

QUIZ TIME!

Q2.5. Risk is individually managed within an IPD. True or False?

   A. True
   B. False

Q2.6. When working on a green building project, when is the best time to incorporate an integrative design approach? (Choose one)

   A. Schematic design
   B. Construction documents
   C. Design development
   D. Beginning of construction
   E. Substantial completion

Q2.7. Life-cycle assessments (LCAs) are a beneficial tool to determine which of the following? (Choose one)

   A. Environmental benefits and potential impacts of a material, product, or technology
   B. Economics of building systems during the life of the building
   C. Environmental impacts of materials during construction
   D. Social impacts of policies during a fiscal year
   E. Maintenance implications, including cost, during the life of the building

Q2.8. The project team is looking to conduct a life-cycle cost analysis as a method of evaluating alternative flooring products. Which of the following should they take into consideration as input factors to that analysis? (Choose two)

   A. First costs, excluding the cost of installation
   B. First costs, including the cost of installation
   C. Maintenance, life expectancy, and replacement cost
   D. Maintenance and replacement cost, but not life expectancy

Q2.9. A LEED Accredited Professional is presented with a project that was started without sustainable design or LEED certification in mind. However, the owner and design team members involved thus far—none of which has significant experience with either LEED or sustainable design—have expressed interest in this path, in spite of the fact that the project is already well under way and moving rapidly toward the construction document phase. Given this situation, which of the following would tend to have the most influence on the effectiveness of the sustainable design process for a project aimed at LEED certification? (Choose three)

    A. Starting the sustainable design process and consideration of LEED-related goals and getting objectives under way as soon as possible

    B. Extensive research, evaluation, and life-cycle assessment for intended material and technology options

    C. Aggressive value engineering of individual line items to ensure that the budget is not exceeded

    D. Collectively delegating responsibility for specific target LEED credits and associated strategies to appropriate team members

    E. Establishing means of collaborative, interdisciplinary communication among team members as a departure from a conventionally more segmented design process

Q2.10. Which of the following are components of green building design and construction projects? (Choose three)

    A. Life-cycle assessments

    B. Schedule increases

    C. Decreases in communication

    D. BIM

    E. IPD

CHAPTER 3
THE LEED FOR COMMERCIAL INTERIORS RATING SYSTEM

CURRENTLY, THE LEADERSHIP IN ENERGY AND ENVIRONMENTAL DESIGN (LEED®) Accredited Professional (AP) Interior Design and Construction (ID+C) exam tests the knowledge of the rating system contained within the *LEED Reference Guide for Green Interior Design and Construction*. Not only is it important to know which rating system is included, but it is also important to know when it is appropriate to pursue the certification, depending on the project type. At the time of printing, the *ID+C Reference Guide* contains the LEED for Commercial Interiors™ (CI) rating system but will eventually also include the LEED for Retail Interiors™ and LEED for Hospitality™ rating systems as well.

LEED FOR COMMERCIAL INTERIORS (LEED CI)

The LEED CI rating system focuses on tenant improvement projects in the commercial market sector primarily aimed at corporate, retail, and institutional project types (Figure 3.1). "Tenants who lease their space or do not occupy the entire building are eligible."[1] Since tenants only occupy a portion of a building,

Figure 3.1 Hotel Felix incorporates many sustainable design strategies such as incorporating low-emitting materials and products with recycled content helping to earn the project a silver certification under the LEED CI rating system. *Photo courtesy of Gettys*

the rating system is intended to work in tandem with the LEED for Core and Shell™ rating system.

As with any LEED rating system, it is up to the project team to determine which system is best suited to their project.

THE CATEGORIES OF LEED

Each of the three LEED rating systems included in the *ID+C Reference Guide* has five main categories:

- Sustainable Sites (SS)
- Water Efficiency (WE)
- Energy & Atmosphere (EA)
- Materials & Resources (MR)
- Indoor Environmental Quality (EQ)

The rating systems also include two other categories that provide bonus points: Innovation in Design (ID) and Regional Priority (RP).

PREREQUISITES AND CREDITS

As shown in Figure 3.2, within each category of each of the rating systems there are prerequisites and credits. For certification, it is critical to remember that

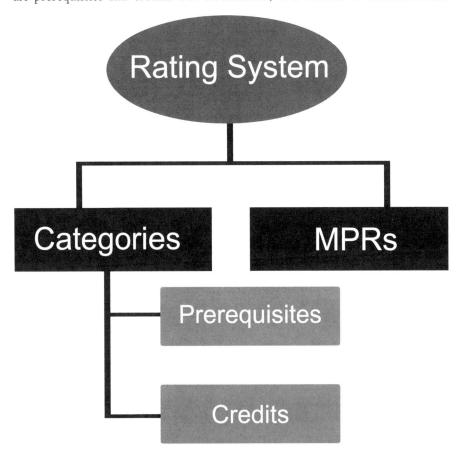

Figure 3.2 The components of a rating system.

prerequisites are absolutely required, while credits are optional. Not all categories contain prerequisites, but all of the categories have credits. All of the prerequisites of each primary category, required by the majority of the rating systems, are noted in the following list (Figure 3.3). These minimum performance features will be discussed in Part II, within each chapter broken out by category. It does not matter if a project intends to pursue credits in every category—*all* prerequisites are required and are mandatory within the rating system the project is working within.

 Make a flashcard to remember the differences between credits and prerequisites. Be sure to include the following: credits are optional components that earn points, whereas prerequisites are mandatory, are not worth any points, and address minimum performance features.

Figure 3.3 The prerequisites covered in the ID+C Reference Guide **are sorted by category:**

Water Efficiency
20 Percent Water Use Reduction

Energy & Atmosphere
Fundamental Commissioning of Building Energy Systems
Minimum Energy Performance
Fundamental Refrigerant Management

Materials & Resources
Storage and Collection of Recyclables

Indoor Environmental Quality
Minimum Indoor Air Quality Performance
Environmental Tobacco Smoke (ETS) Control

Each prerequisite and credit is structured the same way, and both include the same components (Figure 3.4).

Figure 3.4 The components of prerequisites and credits:

- Credit name and point value
- Intent—describes the main goal or benefit for each credit or prerequisite
- Requirements—details the elements to fulfill the prerequisite or credit. Some credits have a selection of options to choose from to earn point(s)
- Benefits and issues to consider—discusses the triple bottom line values to the credit or prerequisite
- Related credits—indicates the trade-offs and synergies of credits and prerequisites
- Referenced standards—lists the standard referenced for establishing the requirements of the credit or prerequisite
- Implementation—suggests strategies and technologies to comply with the requirements of the credit or prerequisite
- Timeline and team—outlines which team member is typically responsible for the credit and when the effort should be addressed
- Calculations—although most calculations are completed online, this section describes the formulas to be used specific to the credit or prerequisite
- Documentation guidelines—describes the necessary documentation requirements to be submitted electronically for certification review
- Examples—demonstrates examples to satisfy requirements

- Exemplary performance—think of these as bonus points for achieving the next incremental level of performance
- Regional variation—speaks to issues as related to project's geographic location
- Operations and maintenance considerations—describes relevance of the credit or prerequisite after building is occupied, specific to the Existing Buildings: Operations & Maintenance rating system
- Resources—provides other tools or suggestions for more information on the topic
- Definitions—provides clarification for general and unique terms presented

Credit Weightings

Remember, prerequisites are not worth any points but are mandatory in order for a project to receive certification.

Refer to the LEED scorecard in Appendix F for a visual representation of how each category is composed of credits and prerequisites and the allocation of points.

Since prerequisites are required, they are not worth any points. All credits however, are worth a minimum of 1 point. Credits are always positive whole numbers, never fractions or negative values. All prerequisites and credits are tallied on scorecards (also referred to as checklists) specific to each rating system.

Any project seeking certification must earn a minimum of 40 points, but this does not mean 40 credits must be awarded as well; because different credits are weighted differently, they have varying point potentials. To determine each credit's weight, USGBC referred to the U.S. Environmental Protection Agency's (EPA's) 13 Tools for the Reduction and Assessment of Chemical and Other Environmental Impacts (TRACI) categories for environmental and health concerns including: climate change, resource depletion, human health criteria, and water intake. Once the categories of impact were determined and prioritized, USGBC referred to the National Institute of Standards and Technology (NIST) for their research to determine a value for each of the credits by comparing each of the strategies to mitigate each of the impacts.

Remember, credit weightings are based on environmental impacts and human benefits, such as energy efficiency and CO_2 reduction for cleaner air.

As a result of the credit weighting and carbon overlay exercise, LEED values those strategies that reduce the impacts on climate change and those with the greatest benefit for indoor environmental quality, focusing on energy efficiency and carbon dioxide (CO_2) reduction strategies. For example, transportation is a very important element within LEED, and therefore any credits associated with getting to and from the project site are weighted more. Water is an invaluable natural resource, and therefore water efficiency and consumption reduction is weighted appropriately to encourage project teams to design accordingly to use less. Providing renewable energy on a project's site will lessen the burden on fossil fuels and, therefore, is also suitably weighted.

In summary, USGBC created a simplified, 100-base-point scale for the four different certification levels.

Make a flashcard so that you can quiz yourself of the certification levels and corresponding point range.

- Certified: 40–49 points
- Silver: 50–59 points
- Gold: 60–79 points
- Platinum: 80 and higher

The 100 base points are totaled from the five main categories: SS, WE, EA, MR, and EQ. The last two categories make up 10 bonus points for a total of 110 available points. Table 3.1 summarizes the point distributions for each of the categories within the LEED CI rating system.

Credits	Category	Possible Points
8	**S**ustainable Sites	24
3	**W**ater Efficiency	11
6	**E**nergy and Atmosphere	33
7	**M**aterials and Resources	13
8	**I**ndoor Environmental Quality	19
	Subtotal	100
Innovation and/or Exemplary Performance		6
Regional Priority		4
	Total	110

Table 3.1 The Point Distributions for the LEED CI Rating System

QUIZ TIME!

Q3.1. A tenant in Arizona is seeking Silver certification for their new space, but are scoring 47 points from the five main categories of the rating system. What should they do? (Choose two)

  A. Nothing, they only need 40 points to earn Silver

  B. Pursue bonus points from the Innovation in Design category

  C. Pursue bonus points from the Regional Priority category

  D. Issue a CIR

  E. Submit an appeal

Q3.2. Which of the following statements are true in regard to credit weightings? (Choose two)

  A. USGBC consulted with NIST and the U.S. EPA's TRACI tool to determine the credit weightings.

  B. The LEED rating systems were reorganized, and new credits were introduced to recognize what matters most, such as transportation.

  C. The LEED CI rating system is based on a 100-point scale.

  D. All credits are worth two points within LEED NC but only one point within LEED CS.

Q3.3. Which rating system is LEED CI meant to work hand in hand with? (Choose one)

  A. LEED for New Construction and Major Renovations

  B. LEED for Core & Shell

  C. LEED for Existing Buildings: Operations & Maintenance

  D. LEED for Retail: Commercial Interiors

  E. LEED for Retail: New Construction and Major Renovations

Q3.4. Which of the following is true about credit weightings and the carbon overlay? (Choose one)

   A. Considers impact of direct energy use

   B. Considers impact of transportation

   C. Considers impact of embodied emissions of water, solid waste, and materials

   D. All of the above

Q3.5. Incorporating green building strategies, such as high-efficiency mechanical systems, on-site photovoltaic systems, and an indoor water wall, to help with the indoor air quality and air conditioning, plays a role in what type of cost implications? (Choose one)

   A. Increased life-cycle costs

   B. Increased first costs

   C. Reduced construction costs

   D. Increased soft costs

   E. Reduced soft costs

Q3.6. If a project plans on earning silver certification under the LEED for Commercial Interiors rating system, which point range would it aim for? (Choose one)

   A. 50–59

   B. 20–30

   C. 40–49

   D. 60–69

   E. 30–39

THE LEED CERTIFICATION PROCESS

The typical project team members were previously outlined, but one member was not included: the LEED project administrator. The administrator is typically responsible for registering a project with the Green Building Certification Institute (GBCI®) and for the coordination of all of the disciplines on the project team, by managing the documentation process until LEED certification is awarded. The LEED project administrator can be one of the team members previously mentioned in Chapter 2 and, therefore, would serve a dual-purpose role, or he or she can be an addition to the team. In either case, the administrator would grant access for each of the team members to LEED-Online, the online project management system.

LEED-Online

TIP: To learn more about LEED-Online, be sure to check out the demo video at www.youtube.com/watch?v=fS3yzjZxcUA.

LEED-Online is a web-based tool used to manage a project seeking LEED certification. It is the starting point to register a project with GBCI and is used to communicate with GBCI for a review of the documentation submitted for both prerequisites and credits during design and construction. All projects seeking certification (except LEED for Homes™) are required to utilize LEED-Online to upload credit forms and any required supporting documentation, such as drawings, contracts, and policies for review by GBCI. Project teams receive reviewer feedback, can check the status of application reviews, and can learn the certification level earned for their project through LEED-Online. Credit interpretation requests and appeals (both to be discussed later in the chapter) are also processed through LEED-Online.

When a team member is invited to a project on LEED-Online, they need to log in to the LEED-Online website to gain access. Once signed in, they are greeted with the My Projects page, where they can see a list of active projects they are assigned to. Upon selecting one of the projects, the Project Dashboard page appears. This dashboard serves as a project's home page and gives access to:

TIP Remember, only invited team members can see a project's LEED-Online page, after a project is registered.

- The project's scorecard, which shows which credits the team is pursuing and their status
- Credit interpretation requests (CIRs) and rulings
- LEED credit forms—think of these as the "cover pages" for each credit and prerequisite. There is a credit form for every prerequisite and credit, which must be submitted through LEED-Online. If a calculation is needed to show compliance, the template contains a spreadsheet with a built-in functionality that automatically completes the calculation after the required data is input according to the requirements described in the reference guides.
- Timeline—where a project administrator would submit for certification review
- Postcertification—to purchase plaques, certificates, and the like

Registration

The LEED certification process for projects begins with project registration. To register a project, the team administrator would sign in to LEED-Online, click the Register New Project tab, and follow the instructions provided. The registration process begins with a review of eligibility criteria, including contact and USGBC membership verification. The next step involves selecting a rating system. LEED-Online provides assistance through a "Rating System Selector" to help the team to decipher which rating system is best suited to the specific project seeking certification. Before advancing to the "Rating System Results" step, the team administrator is prompted to confirm compliance with the seven minimum program requirements (MPRs), which are described in the next section. After confirming MPR compliance, the applicable LEED scorecard appears. The next step of registration includes entering specific project information, including owner contact information, project address and square footage, and the anticipated construction start and end dates. All of the information is then presented on screen for review, the payment is processed, and then the registration information is confirmed. The project administrator is then awarded access to the project's LEED-Online page through the My Projects tab.

TIP Remember, USGBC members pay a reduced fee for project registration and certification.

Minimum Program Requirements

Just as there are prerequisites that must be achieved in each rating system, there are seven MPRs that must be met in order for a project to receive certification. MPRs pertain to all the rating systems except LEED for Homes and LEED for Neighborhood Development. MPRs are critical components that are not listed on a project scorecard, but instead are confirmed when registering a project on LEED-Online. Should noncompliance with any of the seven mandated MPRs be found at any time, a project could risk losing its certification, including any fees paid for registration and certification. The USGBC website details the following seven MPRs[2]:

MPR 1. Must comply with environmental laws

MPR 2. Must be a complete, permanent building or space

MPR 2 prohibits mobile homes, trailers, and boats from pursuing LEED certification.

For a LEED CI project, the "LEED project scope must include a complete interior space distinct from other spaces within the same building with regards to at least one of the following characteristics: ownership, management, lease, or part wall separation."[3]

MPR 3. Must use a reasonable site boundary

Floor separation can serve to define a complete interior space.

The LEED project boundary must include and be consistent with all of the property as part of the scope of work of the new construction or major renovation, including any land that will be disturbed for the purpose of undertaking the LEED project during construction and operations.

The LEED project boundary may not include land that is owned by another party.

MPR 4. Must comply with minimum floor area requirements

For LEED CI projects, a minimum of 250 square feet of gross floor area must be included in the scope of work.

MPR 5. Must comply with minimum occupancy rates

Full-time equivalent occupancy

Remember, all MPRs must be met in order to certify a project and to keep the certification once earned.

The LEED project must be occupied at least one full-time equivalent (FTE) occupant. If there is less than one annualized FTE, IEQ credits will not be awarded, although compliance with IEQ prerequisites is required.

MPR 6. Commitment to share whole-building energy and water usage data

Five years of actual whole-project utility data must be shared with the U.S. Green Building Council (USGBC) and/or Green Building Certification Institute (GBCI).

If it is not practical or is cost-prohibitive to meter the utilities for the entire LEED certified gross floor area, the project may be exempt from complying with this MPR.

MPR 7. Must comply with a minimum building area to site area ratio

The project's gross floor area must be no less than 2 percent of the gross land area within the LEED project boundary.

Be sure to make flashcards to remember the seven MPRs.

MPR 3 refers to the LEED project boundary. There are three types of boundaries to be aware of for the purposes of LEED: property boundary line, LEED project boundary line, and the building footprint (Figure 3.5). The property boundary line refers to the land owned according to a plot plan or legal property deed. The LEED project boundary line may or may not be the same as the property boundary. For example, a university may own acres of land but may wish

Make a flashcard to remember the three types of boundaries associated with LEED projects.

THE LEED CERTIFICATION PROCESS 29

Figure 3.5 The different types of boundaries for a LEED project.

to develop only a portion of it for one academic building. Therefore, the LEED project boundary line sets the limits for the scope of work to be included in the documents for certification. The **building footprint** is the amount of land on which the building resides.

Tenant improvement projects seeking LEED CI certification will most likely have a LEED project boundary line consistent with a building's floorplate or smaller, depending on the amount of space leased. If this is the case, the exterior grounds would not be included and, therefore, the project would be exempt from complying with MPR 7.

Local municipalities are responsible for establishing sections of their towns for different uses and then categorizing these areas into different zones, such as commercial, residential, and industrial. These sections of land are regulated according to:

- Building type (commercial, residential, mixed-use, etc.)
- Building height
- Footprint, impervious versus pervious
- Setbacks
- Parking
- Open space

Project teams should be mindful of local regulations for land and the allowable uses. For the purposes of the exam, it is important to remember that although many credits may reference zoning, LEED will never override local, state, or federal requirements.

TIP Be sure to check out Appendix B for an MPR summary chart.

Credit Form

After a project is registered, the project administrator invites the other team members to the project's LEED-Online site and assigns each member the coordinating prerequisites and credits they will be responsible for. This means each prerequisite and credit has one responsible party assigned to it, and that person

TIP Remember, a credit form acts as a cover page of sorts.

One team member can be assigned to more than one credit or prerequisite. Additional team members can be invited to LEED-Online and not be assigned any credit or prerequisite.

Make a flashcard to remember LPE!

When a CIR is responded to by GBCI, the reply is referred to as a **credit interpretation ruling**.

Be sure to check out the CIR guidelines on the GBCI website at www.gbci.org/Libraries/Credential_Exam_References/Guidelines-for-CIR-Customers.sflb.ashx.

Make a flashcard to remember the details of submitting a CIR.

will generate and upload the required documentation specific to each prerequisite and credit.

When a team member is assigned a credit or prerequisite, he or she becomes the **declarant** for signing the credit form. Each of these forms summarizes how the project team has satisfied the requirements for the specific credit or prerequisite. There is a credit form for every prerequisite and credit, which must be submitted through LEED-Online. If a calculation is needed to show compliance, the form contains a spreadsheet and automatically completes the calculation after the required data is input according to the requirements described in the reference guides. It is suggested that you visit the USGBC website to download the sample credit forms (the link is available through the candidate handbook from GBCI).

Remember, all prerequisites and credits require a credit form, and some may require additional documentation. The additional documentation may be exempt if the design team opts to use the licensed-professional exemption (LPE) path. This optional path is determined on the credit form.

Credit Interpretation Requests and Rulings

For those in the design and construction industry, it is helpful to think of **credit interpretation requests** (CIRs) like a request for information (RFI). Just as a contractor may issue an RFI to the design team for clarification about a detail to be constructed, a project team can issue a CIR to GBCI in an effort to obtain clarification or more information. CIRs can be submitted any time after a project is registered.

For a fee, team members of registered projects seeking LEED certification can submit CIRs for clarification about a credit or prerequisite within a LEED rating system. It is important to remember that CIRs are issued specific to *one* credit, prerequisite, or MPR. Note that CIR rulings are not considered final, nor are they definitive in determining if a particular strategy is satisfactory. Therefore, project teams are encouraged to upload their ruling with the coordinating credit, prerequisite, or MPR when submitting for a certification review through LEED-Online.

The USGBC website contains a database on previously issued CIRs for teams to query for more information before submitting a new one. Although these CIR postings serve as useful tools for a project team, they will no longer be used as supporting documentation for projects other than the one that submitted the CIR. For projects registered under the 2009, version 3 rating systems, CIRs are project specific and will not be posted to the database. With this change, teams are still encouraged to refer to the reference guides and the CIR database, as well as to contact GBCI before submitting a CIR, since there is a fee associated with issuing a new CIR.

Certification Review

Once the design team moves through the design phase and completes the construction documents, they are allowed to submit an application for a design review, although this is optional and not required. Optional split reviews provide the team with a preliminary status to see where the project stands with regard to point-earning potential (at least from a design prerequisite and credit standpoint). If the team decides not to pursue this preliminary review at the end of design, they will wait to submit their documentation until after substantial

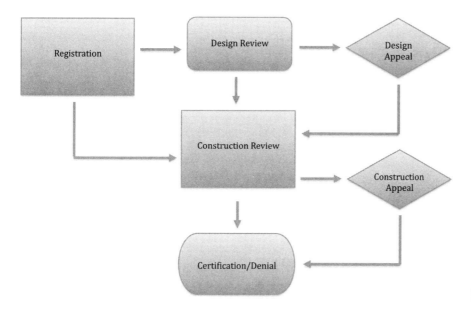

Figure 3.6 The certification review process.

completion. In summary, a project team can choose a split review for two certification reviews (one at the end of the design phase and another at the end of construction) or submit all documentation for a construction review after substantial completion (Figure 3.6).

At the time of a review submission, the LEED project administrator will need to pay a fee for certification review. This certification fee is based on the rating system the project is seeking certification with, the project's square footage, and whether the project was registered under a corporate membership account with USGBC. The project administrator will be required to submit a short project narrative to provide GBCI with the background of the project, the intended use of the project, the location and surrounding areas, and any other details deemed appropriate for clarification purposes. Project photographs or renderings, elevations, floor plans, and any details should also be uploaded to LEED-Online.

 Make a flashcard to remember the four components on which a project's certification fees are based.

Table 3.2 Certification Review Schedule

Process	Days
Preliminary design review or construction review	25 business days
Team reply	25 business days
Final review	15 business days
Team reply	15 business days
Appeal (if necessary)	25 business days
Appeal review (if necessary)	25 business days
Certification awarded	

The Time Frames of Certification Reviews

Table 3.2 outlines the schedule associated with the submission and review times during the LEED certification review process. Bear in mind that the time frames listed apply to both design reviews and construction reviews. Once a project team submits for either type of review, they must then wait 25 business days to hear back from GBCI. GBCI updates LEED-Online to indicate whether a credit or prerequisite is "anticipated" or "denied," or will issue clarification requests to the team specific to any credits or prerequisites in question. The team then has 25 business days to respond with more information to explain how they satisfy the requirements of the credit or prerequisite requiring clarification. At that time, the team must wait 15 business days to receive an indication as to whether the credit or prerequisite clarified is either "anticipated" or "denied." If the team submitted for a design review, they would repeat the steps listed earlier at the end of substantial completion in order to submit for a construction review. It is not until this review process that the final decision to award or deny credits and prerequisites takes place.

Should the team receive a "denied" status, they can issue an appeal to GBCI for a fee within 25 business days. GBCI would then have 25 business days to review the appeal and issue a ruling to the project team. Once the appeal process ends, certification is awarded.

TIP Remember, GBCI is responsible for the appeals process!

QUIZ TIME!

Q3.7. Credit interpretation requests (CIRs) provide which of the following? (Choose two)

   A. Responses to written requests for interpretation of credit requirements

   B. Determination of whether a particular strategy can be used to satisfy two different credits at once

   C. Clarification of one existing LEED credit or prerequisite

   D. Definitive assurance that a particular method or strategy permitted on a previous project will be applicable to other projects in the future

Q3.8. Which of the following meets the MPR regarding minimum gross floor area for LEED for Commercial Interiors rating system? (Choose one)

   A. 250 square feet

   B. 500 square feet

   C. 1,000 square feet

   D. 2,500 square feet

Q3.9. Which of the following statements are not true regarding MPRs? (Choose two)

   A. The LEED project boundary must include all contiguous land that is associated with normal building operations for the LEED project building, including all land that was or will be disturbed for the purpose of undertaking the LEED project.

B. The owner must commit to sharing whole-building energy and water usage data for 10 years.

C. LEED projects located on a campus must have project boundaries such that if all the buildings on campus become LEED certified, then 100 percent of the gross land area on the campus would be included within a LEED boundary.

D. Any given parcel of real property may only be attributed to a single LEED project building unreasonable shapes for the sole purpose of complying with prerequisites.

E. Gerrymandering of a LEED project boundary is allowed.

Q3.10. How many years must a project commit to sharing whole-building energy and water-use? (Choose one)

A. 7 years

B. 10 years, unless the building changes ownership

C. 6 months

D. 1 year

E. 5 years

Q3.11. Which of the following statements are true? (Choose two)

A. Appeals are mailed to GBCI.

B. Appeals can be submitted within 25 business days after final certification review.

C. Appeals are free.

D. Appeals can pertain only to credits, not prerequisites or MPRs.

E. Appeals are submitted through LEED-Online.

Q3.12. Which of the following is new to all LEED rating systems under LEED 2009? (Choose two)

A. Minimum program requirements

B. Credit interpretation requests

C. Regional Priority category

D. Awareness and Education category

E. Credit weighting

Q3.13. Design reviews can prove to be a beneficial option for a team to pursue since the project can be awarded points before construction begins. True or False?

A. True

B. False

Q3.14. LEED project registration provides which of the following? (Choose two)

A. Three credit interpretation requests (CIRs)

B. One pre-application USGBC review of project submittals and documentation

C. One point toward LEED certification for registration prior to the development of construction documents

D. Access to online LEED credit forms for the project

E. Establishment of contact with GBCI

Q3.15. An application for LEED certification must contain which of the following? (Choose two)

A. Project summary information, including project contact, project type, project cost, project size, number of occupants, estimated date of occupancy, etc.

B. A list of all members of the design and construction team, including contact information, documented green building industry experience, and indication of all LEED Accredited Professionals

C. Completed LEED credit forms for all prerequisites and attempted credits, plus any documentation specifically required to support those templates

D. Detailed documentation for all credits pursued, including full-sized plans and drawings, photocopies of invoices for all purchased materials, records of tipping fees, all energy modeling inputs and assumptions, and evidence of all calculations performed in support of LEED credits

Q3.16. When should construction credits and prerequisites be submitted for certification review? (Choose one)

A. Beginning of construction

B. One year after occupancy

C. Substantial completion

D. Once permit is obtained

E. Six months after occupancy

Q3.17. Regarding the application process for LEED certification, which of the following is a correct statement? (Choose one)

A. LEED credit forms and documentation may be submitted only after construction is complete.

B. All LEED credit forms and documentation must be submitted prior to construction.

C. Prerequisites and credits marked as "Design" may be submitted and reviewed at the end of the design phase.

D. The optional design-phase submittal allows projects to secure points for specific LEED credits, for which a preliminary certification will be awarded if the project has earned a sufficient number of points.

Q3.18. How long does a project team have to submit an appeal after receiving certification review comments back from GBCI? (Choose one)

   A. 15 business days
   B. 25 business days
   C. 45 business days
   D. 1 week
   E. 1 month

Q3.19. Which of the following correctly characterize credit interpretation requests (CIRs)? (Choose three)

   A. Can be viewed only by the primary contact for a registered project
   B. Can be submitted any time after a project is registered
   C. Must be requested through LEED-Online
   D. Can be requested only in a written request mailed to GBCI
   E. Can address more than one credit or prerequisite
   F. Are relevant to one specific project and will not be referenced in the CIR database

Q3.20. Who is responsible for the appeals process? (Choose one)

   A. GBCI
   B. USGBC
   C. Tenant
   D. Owner
   E. Project Team Administrator

PART II

DIVING IN: THE STRATEGIES AND TECHNOLOGIES OF LEED CI

CHAPTER 4
SUSTAINABLE SITES

THIS CHAPTER BEGINS THE DETAILED STUDY OF THE INTENTIONS, requirements, and strategies described within the Sustainable Sites (SS) category's credits, as there are no prerequisites. The main topics include the factors applicable to site selection and location, design, construction, and maintenance of the site. As with making any other decision while working on a green building project, all components within the SS category are weighed on the triple bottom line values of environmental, economic, and community impacts.

 Dedicate one color flashcard for all your flashcards within the SS category. This way anytime you see that color, you will associate that flashcard question with Sustainable Sites concepts and strategies.

Where a tenant space is located and how the base building was developed can have multiple impacts on the ecosystem and water resources required during the life of a building. The site location can affect a building's energy performance with respect to orientation; it also could affect stormwater runoff strategies and the potential to pollute nearby waterways. How does the site fit into the existing infrastructure? Was it a contaminated site that was remediated for redevelopment? Carbon emissions should be evaluated as well, as they may be affected by the transportation required to get to and from the site. Is there public transportation access? How much parking is available for cars? Is the project site dependent on the use of cars? If so, are there incentives for carpools or vanpools? These concepts and questions should trigger some of the important factors to consider when deciding on a particular base building and its true sustainable value.

Before a tenant space is selected, the project team needs to evaluate the site and base building for compliance with the Leadership in Energy and Environmental Design (LEED®) for Commercial Interiors™ (CI) rating system. They need to inquire to find out if the base building was previously certified and if not, determine if sustainable strategies were implemented that can help satisfy the requirements to achieve points under the LEED CI rating system. Although the base building can have an impact on strategies for compliance in other categories, the SS category focuses on the concepts achieved outside of the tenant space.

The SS category is composed of the credits shown in Table 4.1. Notice the category does not include any prerequisites. Take note of the points available for

Table 4.1 The Sustainable Sites Category

D/C	Prereq/Credit	Title	Points
D	Credit 1	Site Selection	1 to 5
D	Credit 2	Development Density and Community Connectivity	6
D	Credit 3.1	Alternative Transportation, Public Transportation Access	6
D	Credit 3.2	Alternative Transportation, Bicycle Storage and Changing Rooms	2
D	Credit 3.3	Alternative Transportation, Parking Availability	2

each credit as the SS category ranks second highest, after the Energy and Atmosphere category, for the most amount of point potential. Each credit is eligible for a design (D) side review submission after the construction documents are completed. Use the table as a summary of the category, but for the purposes of the exam and an attempt to break down the information to remember, the category information is presented in this chapter by the overlying category strategies instead of a credit-by-credit approach as seen in the reference guide. Therefore, when preparing for the exam, it is helpful to remember that the SS category is broken down into the following two factors and the coordinating credits:

1. Site selection
2. Transportation

TIP Remember to use the study worksheets at the end of the chapter to remember the details of each credit, such as the intent, requirements, and documentation and calculation requirements.

SITE SELECTION

Determining the location of a green tenant space should be influenced by urban locations to discourage sprawling development into the suburbs where undeveloped (greenfield) sites would then be disturbed. The idea is to build up and not out to help increase density and reduce the negative environmental impacts of building on existing, cohesive natural habitats (Figure 4.1). Some municipalities offer an increased floor-to-area ratio (FAR) incentive to encourage the development of green buildings within certain communities. Zoning departments typically define building setback development lines based on the use and location of a neighborhood or town. Developers with an increased FAR allowance can build more within the setback lines and then, in turn, can sell or rent more space.

From an environmental perspective, the selection of a sustainable site would avoid the development of greenfields and the destruction of wildlife habitats to help lessen the threat to the wildlife's ability to survive. The goal is to preserve

Figure 4.1 Urban development in downtown Miami, Florida.

land and, thus, preserve plant and animal species. From an economic standpoint, avoiding sprawl development helps to lessen the burden of expanding infrastructure for both utilities and transportation. The social equity of the proper site selection could include the protection of the natural environment to be enjoyed and observed by future generations for ecological and recreational purposes.

Site Selection in Relation to LEED Compliance

Project teams are encouraged to address the following two credits when selecting a tenant space, according to the LEED CI rating system.

SS Credit 1: Site Selection. This credit offers teams many options to collect up to 5 points. Projects can quickly earn 5 points by leasing space in a LEED certified base building under the LEED for Core & Shell™ (CS) rating system or any other LEED rating system. Should the building not be certified but has sustainable and environmental features, project teams should evaluate the following 12 compliance paths to pursue. Most of the 12 paths offer one point for complying with the requirements as described below, except for Paths 7–9 and 11, where teams can earn up to 2 points each.

Path 1: Brownfield Redevelopment. There are over 450,000 brownfield remediation and redevelopment opportunities in the United States available to help the reduction of sprawl developments and reuse land, thereby protecting the environment and undeveloped property.[1] These opportunities help to explain the long-term goals of USGBC to encourage regenerative projects. As a result of this goal, the LEED CI rating system helps to encourage the **remediation** and redevelopment of brownfield sites by offering one point should a tenant lease space within a building on a site classified as a Brownfield by a local, state, or federal government agency. The site would have needed to be remediated in order to qualify (Figure 4.2).

 If less parking were provided because urban redevelopment projects had access to public transportation, construction costs would be lower.

 TIP Note that all of the following strategies are addressed during the **Predesign phase.**

 TIP Make a flashcard to remember the EPA's definition of a **brownfield:** "real property, the expansion, redevelopment, or reuse of which may be complicated by the presence or potential presence of a hazardous substance, pollutant, or contaminant. Cleaning up and reinvesting in these properties protects the environment, reduces blight, and takes development pressures off green spaces and working lands."[2]

Figure 4.2 Villa Montgomery Apartments, a remediation project in Redwood City, California, by Fisher-Friedman Associates, earned LEED Gold certification, for compliance with multiple Sustainable Sites strategies, including brownfield site selection and remediation.
Photo courtesy of FFA

Stormwater Management

Stormwater runoff causes degradation of the surface water quality and reduces groundwater recharge to the local **aquifer,** an underground source of water for groundwater, wells, and springs. The reduction in surface water quality is caused by both a filtration decrease and the increase of hardscape areas containing contaminants. The increase of impervious surfaces and stormwater runoff has put water quality, aquatic life, and recreational areas at risk.

Non-point-source pollutants, such as oil leaked from cars or fertilizers from plantings, are one of the biggest risks to the quality of surface water and aquatic life. These pollutants typically contaminate rainwater flowing along **impervious** surfaces on the journey to sewer systems or water bodies, especially after a heavy rainfall. Once this polluted rainwater is in the sewer system, it then contaminates the rest of the water and takes a toll on the process to purify it or it can contaminate the body of water into which it is dumped. These bodies of water also then suffer from soil erosion and sedimentation deteriorating aquatic life and recreational opportunities. Therefore, allowing rainwater to percolate through vegetation (Figure 4.3) or **pervious** surfaces, such as pervious concrete, porous pavement, or open-grid pavers that allows at least 50 percent of water to seep through, reduces the pollution of surface water and is less of a burden on our ecosystem. For projects located in urban areas where space is limited, oil separators can be utilized to remove oil, sediment, and floatables. Within an oil separator, heavier solid materials settle to the bottom while floatable oil and grease rise to the top.

The triple bottom line benefits of managing stormwater include preserving the natural ecological systems that promote **biodiversity**, which in turn help manage stormwater, such as wetlands (Figure 4.4). If the natural environment could manage stormwater, we could take advantage of the economic savings from not having to create manmade structures to do it for us, as well as not having to pay the costs to maintain the structures. There is also social equity in managing runoff and maintaining clean surface water: the preservation of aquatic life and the ability to enjoy recreational activities. With the importance of these triple bottom line benefits, LEED addresses stormwater management within the following two paths under SS Credit 1:

 Create another flashcard to remember the definition of **stormwater runoff**: rainwater that leaves a project site flowing along parking lots and roadways, traveling to sewer systems and water bodies.

 TIP Non-point-source pollutants are one of the biggest risks to the quality of surface water and aquatic life.

 Create a flashcard to remember the definitions of **impervious surfaces**, surfaces that do not allow 50 percent of water to pass through them, and **pervious surfaces**, surfaces that allow at least 50 percent of water to percolate or penetrate through them.

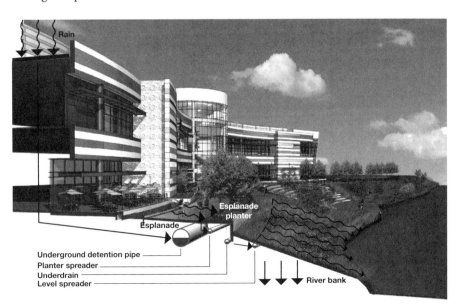

Figure 4.3 Rivers Casino Riverfront Park in Pittsburgh, Pennsylvania, addresses stormwater quantity and quality by incorporating filtration measures to capture and clean rainwater before releasing it into the nearby river.
Photo courtesy of Strada Architecture LLC

Figure 4.4 The Utah Botanical Center's Wetland Discovery Point project, at Utah State University in Kaysville, earned LEED Platinum certification for its efforts to create biodiversity.
Photo courtesy of Gary Neuenswander, Utah Agricultural Experiment Station

Path 2: Stormwater Design, Quantity Control. Project teams will need to determine the predevelopment imperviousness in order to determine which of the two compliance paths to pursue for this credit. If the existing imperviousness was less than 50 percent for the total site prior to development, then the team will be required to determine if the site has increased the imperviousness. If the existing imperviousness was more than 50 percent prior to development, the team will need to ensure that the current stormwater management strategies decrease the runoff volume by 25 percent from the 1 1/2-year, 24-hour design storm. When calculating stormwater values, project teams should use actual local rainfall unless the actual amount exceeds the 10-year annual average local rainfall, in which case the 10-year annual average should be used.

Project teams have a number of strategies from which they can choose to reduce the amount of stormwater that leaves a project site, such as **wet** or **dry retention ponds.** Where dry ponds are only utilized during storms, some wet ponds can permanently hold water. Both of these approaches utilize excavated areas used to detain rainwater from leaving the site and, therefore, slow runoff. Three other options include on-site filtration methods. Bioswales (Figure 4.5), or engineered basins with vegetation, can be utilized to increase groundwater recharge and reduce peak stormwater runoff. The other options, vegetated filter strips and rain gardens; both function to collect and filter runoff while reducing peak discharge rates. Rooftops also contribute to the pollution of surface water, so implementing a green, or vegetated, roof would also reduce stormwater runoff. Another option is to collect the water from the roof and use it for irrigation purposes (Figure 4.6) or toilet flushing.

Path 3: Stormwater Design, Quality Control. Project teams must ensure that the site is capable of removing 80 percent of the **total suspended solids** (TSS) load and 40 percent of the average annual site area's total phosphorous (TP). Total suspended solids are particles that do not settle with gravity, but instead continue to move with the flow of stormwater and, therefore, need to be removed with filters instead to reduce contamination. Strategies to comply with this compliance path include vegetated swales, green roofs (Figure 4.7), and pervious paving. Should

 Green roofs have many synergies, including maximizing open space, creating a habitat for wildlife, and reducing stormwater runoff and the heat island effect, but they also help to insulate a building and, therefore, reduce energy use. The trade-offs with green roofs include installation cost and maintenance. Remembering these types of synergies is the key to success on the LEED AP ID+C exam

 Reducing the amount of impervious surfaces, and increasing pervious surfaces, helps to reduce stormwater runoff and, therefore, also helps to preserve the quality of water.

Figure 4.5 Bioswales, an on-site filtration strategy, can help to recharge the groundwater and reduce stormwater runoff.
Photo courtesy of Thomas M. Robinson, EDR

a site not meet the requirements of this compliance path, the site will need to be modified in order to earn this point, as with all of the strategies of SS Credit 1.

Heat Island Effect

Although energy use will be discussed in more detail in Chapter 6, materials used for site design and rooftops can efficiently impact the use of energy for two reasons. Think of summertime at the grocery store parking lot and how you can see heat emitting from the black asphalt surface. The sun is attracted to darker surfaces, where heat is then retained. Multiply this effect in a downtown, urban area to truly understand the impacts of the urban **heat island effect**. By specifying and implementing materials with a high **solar reflectance index (SRI),** green building projects can reduce the heat island effect and the overall temperature of an area. A material's SRI value is based on the material's ability to reflect or reject solar heat gain measured on a scale from 0 (dark, most absorptive) to 100 (light, most

 Besides low-albedo, nonreflective surface materials, car exhaust, air conditioners, and street equipment contribute to the heat island effect, while narrow streets and tall buildings make it worse.

 Create another flashcard to remember the definition of **heat island effect:** heat absorption by low-SRI hardscape materials that contribute to an overall increase in temperature by radiating heat.

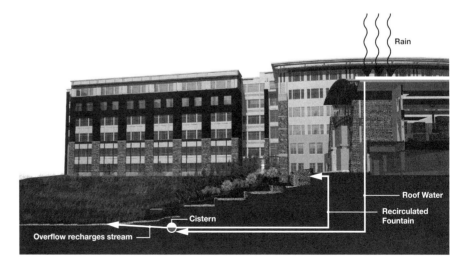

Figure 4.6 Dick Sporting Goods corporate headquarters in Pittsburgh, Pennsylvania, designed by Strada Architecture, LLC, incorporates roof water capture to reduce the stormwater runoff and reduce the need for potable water for irrigation.
Photo courtesy of Strada Architecture LLC

Figure 4.7 This green roof at the Allegany County Human Resources Development Commission's community center in Cumberland, Maryland, minimizes impervious areas to reduce stormwater runoff. *Photo courtesy of Moshier Studio*

reflective). Building materials are also evaluated based on their ability to reflect sunlight based on visible, infrared, and ultraviolet wavelengths on a scale from 0 to 1 (Figure 4.8). This solar reflectance is referred to as **albedo;** therefore, the terms *SRI* and *albedo* can be thought of synonymously for the purposes of the exam.

Path 4: Heat Island Effect, Nonroof. Project teams have three options from which to choose to comply with this compliance path to reduce the heat island effect. Option 1 addresses the hardscape, option 2 impacts a parking provision strategy, and option 3 is a combination approach. Teams have the option to select a base building with a combination of shading (if not existing, within five years), high-SRI materials (minimum of 29), and/or open-grid paving (Figure 4.9) for

 It is the collective strategies that have the biggest reduction of building-associated environmental impacts, such as tuck-under parking with reserved spaces for low-emitting fuel types and carpools/vanpools.

 Create two more flashcards to remember the definitions of solar reflectance **index (SRI)** and **albedo**. Also, remember the different scales for each.

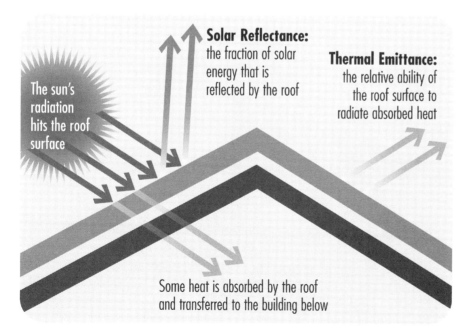

Figure 4.8 Diagram illustrating solar reflectance and thermal (or infrared) emittance. *Image courtesy of Cool Roof Rating Council*

30 percent of the hardscape areas in order to comply with option 1. If high-SRI materials are used for surface paving and walkways, light can be distributed more efficiently at night to reduce the number of light fixtures required, which saves money during construction and later, during operations.

Option 2 requires 50 percent of the parking to be provided under cover. Not only does tuck-under parking help to reduce the impacts of the urban heat island effect, but the strategy also helps to reduce impervious surfaces and helps to preserve open space for parking and nonparking uses. It is important for teams to determine ways to limit the number of impervious surfaces on the site, not only to reduce the heat island effect, but to reduce stormwater runoff as well. For the purposes of the exam, it is important to remember these types of synergies of strategies and credits.

Option 3 requires at least 50 percent of the parking area to utilize an open-grid pavement system (with less than 50 percent imperviousness).

If the team were seeking an exemplary performance point under Path 12, they would need to prove compliance with at least two of the three compliance options.

 Create a flashcard to remember the examples of nonroof impervious surfaces, such as parking areas, walkways, plazas, and firelanes.

Figure 4.9 Turfstone™ Open-Grid Pavers allow stormwater to pass through and encourage vegetation growth between each open cell, in order to recharge groundwater and reduce runoff. *Photo courtesy of Ideal*

SITE SELECTION 47

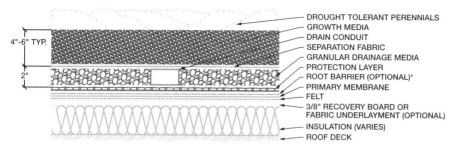

Figure 4.10 Assessing the different components of a green roof serves as a reminder of all of the different team members required to develop an appropriate solution.
Image courtesy of Roofscapes, Inc.

Path 5: Heat Island Effect, Roof. Project teams have three options to choose from in order to pursue this point. The first addresses the roofing material. If a lighter roofing material is used, the mechanical systems do not have to compensate for the heat gain to cool a building, consequently reducing the use of energy. Therefore, if at least 75 percent of the roof surface (not including equipment, photovoltaic solar panels, and penetrations) is covered with a material with high SRI, then the project team could be awarded this credit and earn one point. If the roof is low-sloped, the minimum required SRI is 78, while a steep-sloped roof can comply with a material with an SRI of at least 29. Option 2 requires at least 50 percent of the roof surface to be covered with vegetation (Figure 4.10). Option three allows the team to provide a combination of options 1 and 2 (Figure 4.11).

 Create a flashcard to remember the definition of **emissivity** as described in the ID+C Reference Guide: "the ratio of the radiation emitted by a surface to the radiation emitted by a black body at the same temperature."[3]

 TIP Notice that a minimum of SRI 29 is used for both Paths 4 and 5!

 TIP Create a flashcard to remember steep-sloped roof is considered to have a pitch of greater than 2:12 while a low-sloped roof has a slope of greater than or equal to 2:12.

Figure 4.11 The roof at Villa Montgomery Apartments in Redwood City, California, reduces the urban heat island effect by the installation of a combination of a high-SRI roof material with a photovoltaic system (generating a portion of the electricity needed for operations), along with a green roof, including a playground that offers residents the opportunity to enjoy the outdoor environment.
Photo courtesy of FFA

If a project team doubles compliance for either option, they can pursue an exemplary performance point under Path 12.

Path 6: Light Pollution Reduction. "**Light pollution** is waste light from building sites that produces glare, is directed upward to the sky, or is directed off the site."[4] Project teams will need to address exterior lighting, as well as interior lighting for this design credit. In terms of site lighting, traditionally there has been little attention paid to the quality of the night sky and the effects on wildlife, or to the wasteful energy use approach for exterior lighting. It is inefficient to illuminate areas not used at night, light areas beyond a property's boundary, or overcompensate light levels, or **footcandle** levels. If vertical footcandle levels are minimized, light pollution is reduced, dark night skies are preserved, visibility at night is improved by the reduction of glare, and nocturnal animal habitats remain unaffected from sky glow.

Create a flashcard to remember the definition of light pollution.

Buildings that operate 24 hours per day are required to comply with Option 2 of Path 6: Light Pollution Reduction.

Project teams have two options from which to choose to earn this point. Option 1 requires the tenant space to be located in a building with nonemergency interior light fixtures that are automatically controlled to turn off (or a 50 percent reduced input power between the hours of 11 P.M. and 5 A.M. Option 2 requires the tenant space to be located in a building with shading devices at the exterior openings (Figure 4.12). The shading devices need to automatically close between the hours of 11 P.M. and 5 A.M. and have a transmittance of less than 10 percent. Override switches are allowed with a 30-minute default.

Create a flashcard to remember the definition of a **footcandle**: a measurement of light calculated in lumens per square foot.

Project teams have two options for complying with the interior light component of this credit. At any envelope openings, they can either provide a type of control to reduce the interior lighting power by 50 percent or they can shield all the nonemergency light fixtures with a controlled mechanism, such as automatic shades with a transmittance of less than 10 percent. Either option is to be provided following regular business hours, or between the hours of 11 P.M. and 5 A.M.

Light trespass is unwanted light shining on another's property and can be distracting causing annoyance, discomfort, and loss of visibility.

Water Efficient Landscaping

Irrigating landscaping is the primary use of outdoor water, and is, therefore, a component to be addressed and reduced for projects seeking LEED certification. Site design, including **native (indigenous) and adaptive (introduced) plants** (Figure 4.13) can drastically reduce the amount of water required for irrigation, if not eliminate the need for irrigation all together. If irrigation is required, implementing a high-efficient system can also substantially reduce the amount of water required over conventional designs. Green building projects might also implement another sustainable option, including capturing rainwater to use for irrigation and indoor water flush functions.

Remember, both composting and mulching optimize soil conditions to add to the efficiencies of native and adaptive plantings and high-efficiency irrigation systems.

Native plantings refer to native vegetation that occurs naturally; while **adaptive plantings** are not natural, they can adapt to their new surroundings. Both can survive with little to no human interaction or resources. Project teams should select plants that will not only require less water and maintenance but also improve the nutrients in the soil and deter pests at the same time. Implementing these measures reduces the amount of chemicals into the water infrastructure and, therefore, improves the quality of surface water and saves building owners from purchasing fertilizers and pesticides. Avoiding or reducing the amount of **potable water** (drinking water supplied by municipalities or wells) used for irrigation decreases the quantity of water required for building sites and, therefore, reduces maintenance costs as well.

Create flashcards to remember each of the definitions for **native and adaptive plantings** and **potable water**.

Figure 4.12 Providing glare control components, such as these interior roller shades, can help to satisfy light pollution reduction strategies.
Photo courtesy of Lutron Electronics Co., Inc.

The heat island effect and how it is responsible for an overall temperature increase of an area were introduced earlier. The combination of greenhouse gas emissions, the heat island effect, and increased impervious surfaces from sprawling developments, is resulting in water evaporating at quicker rates and not getting delivered to plants and vegetation. Project teams need to be aware of these conditions and plan accordingly, efficiently, and sustainably. To calculate the amount of water required for the particular types of vegetation and how much of that water is actually delivered and not blown away or evaporated, a project team evaluates the irrigation system with factors such as: total landscape

Figure 4.13 Native and noninvasive plantings do not require irrigation or fertilizers.

 The amount of water delivered by sprinkler heads is measured in gallons per minute (gpm).

 If a project team is required to use potable water for irrigation for public health reasons, the mandate does not hinder achieving this credit, as those areas can be excluded from the calculations. An example might include irrigating near a swimming pool.

area; vegetation characteristics such as species factor, density factor, and microclimate factor; evapotranspiration rate, and irrigation system information such as type and controller efficiency. For the purposes of the exam, it is important to remember the factors of certain calculations, and not necessarily how to perform the calculations.

Paths 7 and 8: Water Efficient Landscaping. Project teams have two opportunities to earn up to 4 points within this pair of paths. If the site uses 50 percent less potable water for irrigation purposes, they can earn two points under Path 7. Path 8 offers another two points if the site does not have a permanently installed irrigation system OR if the irrigation system uses collected rainwater. Reducing the demand can be achieved by implementing a high-efficiency irrigation system with moisture sensors included. These irrigation system types include surface drip, underground, and bubbler systems. A team could also reduce the potable

Figure 4.14 Xeriscaping helps to reduce the demand for potable water with the help of a high-efficiency irrigation system and soil improvements for native plantings.

water demand by specifying native and adaptive plantings and xeriscaping (Figure 4.14).

Path 9: Innovative Wastewater Technologies. Project teams have two options to choose from when pursuing this compliance path for 2 points. Option 1 requires the tenant to lease a space within a building that reduces the potable water consumption used for urinals and toilets by 50 percent by using waterless fixtures, low-flow fixtures (Figure 4.15) or **graywater** or stormwater for flushing and to convey sewage.

Remember, stormwater collection reduces the runoff quantities for the site. Therefore, because there are multiple benefits of capturing stormwater, it is important to understand how a project team would design to collect and reuse the water. Project teams need to evaluate different options to determine the appropriate collection systems for their specific project. Systems can range from

TIP Xeriscaping can also be used as a sustainable landscaping strategy, as it uses drought-adaptable and minimal-water plant types along with soil covers, such as composts and mulches, to reduce evaporation.

 Create a flashcard to remember the definition of **graywater**: wastewater from showers, bathtubs, lavatories, and washing machines. This water has not come into contact with toilet waste according to the International Plumbing Code (IPC).

Figure 4.15 Using dual-flush toilets reduces the need for potable water at the Utah Botanical Center's Wetland Discovery Point building.
Photo courtesy of Gary Neuenswander, Utah Agricultural Experiment Station

> **TIP** Wastewater from toilets and urinals is considered **blackwater**. Kitchen sink, shower, and bathtub wastewater is also considered sources of blackwater. Remember, it's not the source that matters, but what could be in it! For example, washing machine wastewater could be considered blackwater, as washing machines are used to wash cloth diapers.

small barrels to large cisterns (Figures 4.16). Regardless of the system desired, project teams are encouraged to evaluate the following:

- *Water budget.* How much precipitation is expected versus how much water is needed for the purpose the water is intended?
- *Drawdown.* How much water is needed in between rainfalls?
- *Drainage area.* How will the water be collected to store? Will it be a permeable surface? If so, what is the size of the surface to determine how much water can be collected?
- *Conveyance system.* Different pipes will be needed as stormwater and graywater pipes are not allowed to be connected to potable water lines.

Figure 4.16 Stormwater is collected on-site and stored in cisterns at the Utah Botanical Center's Wetland Discovery Point building and used to flush toilets, as well as irrigate the site, thus reducing the need for potable water.
Photo courtesy of Gary Neuenswander, Utah Agricultural Experiment Station

- *Pretreatment.* Screen and/or filters will be needed to remove debris from runoff.
- *Pressurization.* A pump maybe required depending on the system.

Option 2 of this credit requires the tenant space to be located in a base building that treats 100 percent of wastewater to **tertiary standards,** the highest form of wastewater treatment. The treated water would then need to be used on-site or infiltrated. This typically requires a large amount of property to do so and, therefore, lends itself to be an option for campus settings. The key is to avoid aquifer contamination as posed by current septic system technology. Qualifying projects typically implement strategies such as "constructed wetlands, mechanical recirculating sand filters, and anaerobic biological treatment reactors"[5] to comply with this compliance path option.

Path 10: Water Use Reduction. The next chapter will discuss water efficient practices and strategies in more detail (Figure 4.17), including calculating compliance (such as occupancy, presented in SS Credit 3.2 later in the chapter). For now, it is important to remember that commercial interior teams can pursue this point if the tenant leased a space within a building that uses 30 percent less potable water for the entire building. Should the building demand 40 percent less water than a conventional building, the team can pursue an exemplary performance point under Path 12.

Renewable Energy

Keeping with the same triple bottom line goals previously discussed, green building projects with renewable energy technologies can reduce the need to produce and consume coal, nuclear power, and oil and natural gases for energy, thus reducing pollutants and emissions, as well as increasing air quality. For the purposes of LEED compliance, eligible renewable energy sources include solar thermal, photovoltaic systems, wind, wave, some biomass systems, geothermal (heating and electric) power, and low-impact hydropower systems. The *ID+C Reference Guide* provides the following compliance path to address renewable energy and reduce the use of fossil fuels for projects seeking LEED certification. Be sure to make note, Chapter 6 will also speak about renewable energy as related to the tenant.

Path 11: On-site Renewable Energy. If the tenant has leased within a building with at least 2.5 percent of the building's total energy use (based on cost) from an on-site eligible renewable energy source (Figure 4.18), the project could earn 1 point. If the supply doubles and 5 percent of the total building's energy use (based on cost) is sourced from an on-site renewable system, the team can pursue another point. Project teams can pursue a third point under Path 12 should the building have 10 percent of its energy supplied by an eligible on-site renewable energy source (based on cost).

Path 12: Other Quantifiable Environmental Performance. This point earning opportunity is unique to the LEED CI rating system as all other rating systems only have one place to collect and log exemplary performance and other quantifiable achievements not stipulated and addressed within the rating system: under Innovation in Design (ID) Credit 1. Refer to Appendix F for sample LEED CI scorecard to learn more about the ID category point opportunities for innovation and exemplary performance. Just like ID Credit 1, this path allows for both types of achievements to be logged but as only related to SS Credit 1. All

 Water-efficiency strategies incorporated into site design, such as collecting rainwater on-site, can help to reduce the demand for indoor water for flush fixtures.

 Collected stormwater can be used for irrigation, fire suppression, flush fixtures, and custodial uses.

 Don't worry too much about what tertiary treatment is, just be sure to remember the requirement specific to this compliance path.

 Make a note, Chapter 5 will discuss how to calculate occupancy, as it is required as part of the compliance documentation for Path 9.

 Create a flashcard to remember the types of qualifying renewable energy sources. Ineligible systems include any architectural features, passive solar strategies, daylighting strategies, and geo-exchange systems using a ground-source heat pump. If it cannot be metered or uses electricity to operate, it cannot achieve this compliance path.

 Project teams that implement an on-site renewable energy system and sell the renewable energy certificates (RECs) are still able to pursue this compliance path.

54 CHAPTER 4: SUSTAINABLE SITES

Figure 4.17 Aerators provide a low-cost solution to conserving potable water.
Photo courtesy of NEOPERL Inc.

 Can you remember which SS Credit 1 compliance paths are eligible for exemplary performance under Path 12? Hint: There are four.

other achievements outside of this credit shall be logged under ID Credit 1 (see Chapter 9 for more information).

For teams looking for guidance on other quantifiable achievements, they can look to other rating systems, such as LEED CS and NC SS Credit 4.3: Alternative Transportation, Low-Emitting & Fuel Efficient Vehicles (Figure 4.19), and outside

Figure 4.18 Stoller Vineyards in Dayton, Oregon, generates electricity on site by the means of photovoltaic panels mounted on the roof.
Photo courtesy of Mike Haverkate, Stoller Vineyards

of the SS category, such as Indoor Environmental Quality (EQ) Prerequisite 3: Minimum Acoustical Performance from the LEED for Schools™ rating system. They also can propose another strategy through the means of a credit interpretation request (CIR).

SS Credit 2: Development Density and Community Connectivity. Remember from the LEED Green Associate exam the importance to increase density, maximize square footage, protect **greenfields,** and minimize impacts on land. When selecting a project site, developers should focus on **development density** or community connectivity concepts and to select a site with existing utility access (Figure 4.20). To comply, project teams can choose to document that the previously developed site is within an area with a density of 60,000 square feet per acre net or is within a half mile of a residential neighborhood with an average density of 10 units per acre. The latter approach, referred to as **community connectivity**, requires pedestrian access to at least 10 basic services from the new development. These basic services include local businesses and community services such as parks, grocery stores, banks, cleaners, pharmacies, and restaurants. Two of the basic services can be anticipated but would only qualify if they are intended to be operational within one year. Restaurants are the only type of basic service that can be counted twice. For example, if there were five restaurants, three banks, and two grocery stores within a half-mile of the project accessible by pedestrians, then two of the restaurants, one of the banks, and one of the grocery stores would count toward achieving this credit. The team would need to research the area to determine if there are at least six more basic services within the compliance area to achieve option 2 of this credit. For mixed-use projects, only one of the basic services located within the LEED project boundary may count toward the minimum 10 required for compliance. Complying with either development density or community connectivity compliance option earns the team 6 points toward certification.

 Related credits: Selecting an urban project location might help teams to achieve this credit along with SSc3.1: Alternative Transportation, Public Transportation Access.

 Make a flashcard to remember that previously developed sites include not only properties with existing buildings and hardscape but also those that were graded or somehow altered by human activity.

 Make a flashcard to remember the definition of **building density**: the evaluation of a building's total floor area as compared to the total area of the site, measured in square feet per acre.

 TIP Be sure to review the study worksheet at the end of this chapter to remember the three equations associated with the Development Density credit to determine the development density, density radius, and the average property density within density boundary for a project site.

 Make a flashcard to remember the two site selection credits.

CHAPTER 4: SUSTAINABLE SITES

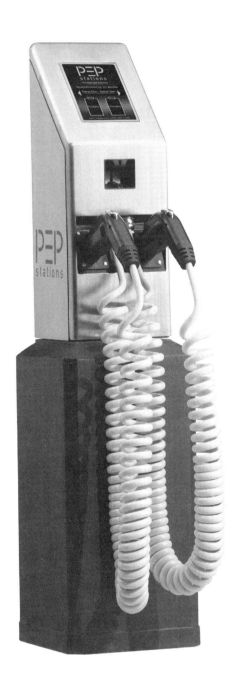

Figure 4.19 Plug-in electric power (PEP) stations help meet the requirements of LEED CS and NC SS Credit 4.3: Alternative Transportation, Low-Emitting & Fuel Efficient Vehicles. The credit requirements can be applied to Path 12 of LEED CI SS Credit 1 to help project teams pick up another point.
Photo courtesy of PEP Station

QUIZ TIME!

Q4.1. What is the foundation for sustainable design for individual buildings? (Choose one)

    A. Carbon emissions

    B. Location

    C. Water use

    D. Orientation

    E. Energy use

Figure 4.20 Gettys Hotel Felix in Chicago, designed by Gettys, successfully earned LEED CI Silver certification; its urban location helped earn the project SS Credits 2 and 3.
Photo courtesy of Gettys

Q4.2. What type of assessment is used to determine brownfield sites? (Choose one)

   A. ASTM C1540 Brownfield Testing
   B. ASTM E1903-97 Phase II Environmental Site Assessment
   C. ASTM E408 Standard Test Methods for Total Emittance of Surfaces, Using Inspection Meter Techniques
   D. ASTM E789 Standard Test Methods for Evaluation for Sites

Q4.3. Which of the following could be a result of sky glow? (Choose three)

    A. Compromised astronomical research

    B. Harm nocturnal environments

    C. Harm diurnal environments

    D. Compromised view of the night sky

    E. Increased view of the night sky

Q4.4. Which of the following represent in-situ remediation solutions for brownfield sites? (Choose two)

    A. Reactive trenches

    B. Pump and treat

    C. Injection wells

    D. Land farming

    E. Bioreactors

Q4.5. Which of the following is the best landscape design strategy to implement to reduce heat island effects? (Choose one)

    A. Absorption

    B. Xeriscaping

    C. Increased albedo

    D. Deciduous trees

    E. Increased imperviousness

Q4.6. Which of the following strategies can help reduce the amount of potable water required for irrigation purposes? (Choose four)

    A. Rainwater collection systems

    B. Planting native, noninvasive vegetation

    C. Hardscape surfaces with an SRI value of at least 29

    D. Using a microirrigation system

    E. Using an sprinkler irrigation system

    F. Using graywater for irrigation

Q4.7. Emissivity is an indication of which of the following material properties? (Choose one)

    A. Ability to reflect light from light sources

    B. Ability of a transparent or translucent material to admit solar gain

    C. Ability of a material to absorb heat across the entire solar spectrum, including all nonvisible wavelengths

    D. Ability of a material to give up heat in the form of long-wave radiation

Q4.8. The design team has attempted to address 50 percent of the hardscape surfaces on the project's site to meet the requirements to reduce the heat island effect for LEED compliance. Which of the following strategies should the LEED Accredited Professional discuss with the design team? (Choose three)

   A. Effective tree-shaded area of hardscape features
   B. Solar reflectance index for all nonstandard paving materials proposed
   C. Percentage of perviousness for proposed open-grid paving materials
   D. Emissivity of all low-albedo hardscape features in the design
   E. Runoff coefficients for impervious paving materials selected

Q4.9. Which of the following statements is not true concerning infrared (or thermal) emittance? (Choose one)

   A. Most building materials have an emittance of 0.9.
   B. Emittance ranges from 0 to 1 (or 0 percent to 100 percent).
   C. Glass is an exception to the rule and has an emittance of 0.1.
   D. Untarnished galvanized steel has low emittance levels.
   E. Aluminum roof coatings have intermediate emittance levels.

Q4.10. An existing base building currently requires 360,791 gallons of water for irrigation for shrubs and turf grass with 180,000 gallons supplied by collected rainwater. The site design incorporated a high efficiency irrigation system including moisture sensors and weather database controllers. In order to calculate the efficiency and reduction in potable water, the team learned a conventional design would require 2,027,825 gallons of potable water. What is the percentage reduction achieved? (Choose one)

   A. 82 percent
   B. 11 percent
   C. 91 percent
   D. 89 percent

Q4.11. What might a project team need to determine to prove compliance with the previous question? (Choose two)

   A. Species factor
   B. SRI values
   C. Evapotranspiration rate
   D. Compliance with ASTM E1903
   E. The 1½ year, 24-hour design storm rate
   F. Imperviousness

Q4.12. What documentation might be required to be uploaded to LEED-Online for Q4.10? (Choose two)

 A. Grounds keeper crew schedule
 B. Landscape plan
 C. Cut- sheets on the rainwater collection system
 D. Local ordinance for irrigation allowances
 E. Design case irrigation system
 F. Baseline case irrigation system

TRANSPORTATION

Transportation is one of the key components addressed within the LEED rating systems as it accounts for 32 percent of the total U.S. greenhouse gas emissions in 2007, according to the U.S. Energy Information Administration.[6] As buildings traditionally have contributed to the need for transportation, green buildings have the opportunity to affect these statistics by reducing the "length and frequency of vehicle trips and encourage shifts to more sustainable modes of transportation."[7] The environmental benefits of sustainable strategies for transportation include a reduction in pollution, including vehicle emissions, which have a dramatic impact on climate change, smog, and acid rain, among other air quality problems, according to Wikipedia.[8] The economic benefits include the reduction of the need to build and maintain roadways. The social component to reducing transportation impacts includes an improvement of human health by increasing the accessibility and therefore encouraging people to walk or bike from place to place.

> **TIP** Land use is the ultimate contributor to the demands of transportation.

Transportation is most impacted by four factors:

- Location—number and frequency of trips
- Vehicle technology—quantity and types of energy and support systems needed to move people and goods to and from the site
- Fuel—environmental impact of vehicle operation
- Human behavior—a daily transportation decision combining the listed impacts[9]

Transportation in Relation to LEED Compliance

The LEED CI rating system offers the following three alternative transportation credits to help reduce the environmental impacts of the use of cars. If more people carpooled, walked, or biked to work or school, there would be less pollution, as vehicles contribute to air pollution, **greenhouse gas** (GHG) emissions, and smog. GHGs "absorb and emit radiation at specific wavelengths within the spectrum of thermal infrared radiation emitted by the Earth's surface, the atmosphere itself, and by clouds."[10] The following alternative transportation strategies help to reduce the use of the automobile and, therefore, help to reduce these detrimental environmental impacts.

Project teams can pursue one Exemplary Performance point for any of the following credits, if a Comprehensive Transportation Management Plan is conducted to prove a measurable reduction in the use of a car to commute to the project site. Remember, moving forward all exemplary performance achievements are documented under ID Credit 1.

SS Credit 3.1: Public Transportation Access. Choose a project site close to **mass transit** (Figure 4.21) to earn 6 points. One or more stops for two or more bus or streetcar lines is required to be within a quarter-mile walk of the building entrance to comply. Projects could also comply if a rail station, ferry terminal, or tram terminal (existing or planned and funded) is located within a half-mile walk of the building's entrance.

In addition to implementing a Comprehensive Transportation Management Plan, as mentioned previously, project teams can also pursue an Exemplary Performance credit if the number of commuter rail, light rail, or subway lines is doubled or if the number of public or campus lines available increases from two to four AND the frequency of service is at least 200 transit rides per day.

SS Credit 3.2: Bicycle Storage and Changing Rooms. Projects seeking LEED certification are encouraged to incorporate bicycle storage (Figure 4.22) and changing rooms to help deter the use of single-occupancy automobiles as a main source of transportation. To comply, project teams will need to first determine the number of bicycle racks to provide by calculating the peak number of occupants, including **transient users.** Next, they will need to calculate the amount of **full-time equivalent (FTE)** occupants to then establish the number of showers and changing rooms to provide. Notice that FTE does not indicate full-time *employee,* but instead full-time *equivalent.* Therefore, to calculate an FTE value, one would include regular building occupants (full-time and part-time) based on an occupancy period of 8 hours. For example, a full-time employee would have an FTE value of 1 because he or she spends 8 hours a week in the

 Selecting a site within an urban context tend to allow project teams to achieve SS Credit 1, SS Credit 2, and SS Credit 4.1.

 Remember, quarter-mile bus or half-mile rail.

 Calculating FTE is required for a few different credits and must be consistent for all of them. Do you remember which compliance path under SS Credit 1, mentioned occupancy calculation? Make note to look for FTE in the next chapter to calculate water use.

 Remember, showers and changing rooms are required to be located within 200 yards of the main building entrance. This could indicate that the facilities can be located in another facility as long as they are accessible and free for tenant occupants.

Figure 4.21 Mass transit in New Jersey is designed to meet the commuting needs of large groups of people.
Photo courtesy of David Cardella

CHAPTER 4: SUSTAINABLE SITES

Figure 4.22 Bike racks at the Animal Care and Protective Services facility in Jacksonville, Florida, provide the staff and visitors the opportunity to commute to the facility without a car.
Photo courtesy of Auld & White Constructors, LLC

building, and 8 divided by 8 is equal to 1. Table 4.2 summarizes the requirements of this credit as project teams would need to know what percentage of the peak occupants and FTE is required for compliance.

Table 4.2 The Requirements of SS Credit 3.2

Bike Racks	Showers/Changing Rooms
5% peak tenant occupants	0.5% FTE occupants

SS Credit 3.3: Parking Capacity. In order to reduce the number of impervious surfaces and encourage the use of mass transit or bicycle commuting, sustainable projects should not have a surplus of parking. Reducing the amount of parking also minimizes the amount of land to be developed, thus lowering construction costs. This alternative transportation credit limits the amount of parking based on the amount of area the tenant occupies.

Case 1 applies to tenants that occupy less than 75 percent of the total building area and offers two different compliance options. With option 1, the tenant parking allotment cannot exceed local zoning ordinances, and 5 percent of the tenant occupants must have preferred parking for carpools or vanpools (Figure 4.23). Option 2 is awarded to tenants with no provided or subsidized parking.

Case 2 applies to tenants that occupy 75 percent or more of the total building area and also offers two different compliance options. With option 1, the tenant parking allotment cannot exceed local zoning ordinances, and 5 percent of the tenant occupants must have preferred parking for carpools or vanpools. Option 2 applies to rehabilitation projects and requires no new parking to be added, with preferred parking for carpools or vanpools serving 5 percent of the tenant occupants.

> **TIP** Make a flashcard to remember the three alternative transportation credits.

Figure 4.23 Providing preferred parking for car/vanpool vehicles could contribute to earning SS Credit 3.3.

QUIZ TIME!

Q4.13. Which of the following alternative transportation credits is affected by site selection? (Choose one)

   A. SS Credit 3.1: Alternative Transportation, Public Transportation Access

   B. SS Credit 3.2: Alternative Transportation, Bicycle Storage and Changing Rooms

   C. SS Credit 3.3: Alternative Transportation, Parking Capacity

   D. None of the above

Q4.14. Which of the following represent the major factors that impact transportation effects on the environment? (Choose three)

   A. Vehicle technology

   B. Fuel

   C. Human behavior

   D. Quality of roads

   E. Suburban development

Q4.15. Which of the following are sustainable strategies that should be implemented on an auto-dependent green building? (Choose three)

A. Provide priority parking for carpools/vanpools.

B. Provide a mass transit discount program to employees.

C. Offer preferred parking rates for multioccupant vehicles.

D. Incorporate basic services (such as a bank, gym, cleaners, or pharmacy) for occupant usage in the new building.

Q4.16. Which of the following are effective and sustainable strategies to address transportation for a LEED project? (Choose three)

A. Choose a site near a bus stop.

B. Limit parking.

C. Encourage carpooling.

D. Provide sport-utility vehicles (SUVs) for all employees.

E. Choose a greenfield site.

Q4.17. Which of the following could contribute to earning SS Credit 1, Path 11: On-site Renewable Energy? (Choose three)

A. Passive solar design concept that captures winter heat from the sun

B. Photovoltaic panels that provide electricity to the building

C. A wind farm located within 500 miles of the project and operated by the local utility company

D. A ground-source heat pump that takes heat from the ground

E. Solar hot water system

F. An on-site electric generator powered by geothermal energy

G. A solar farm adjacent to the project site providing clean power to the grid

Q4.18. Which type of project is required to comply with Option 2 if pursuing SS Credit 1, Path 6: Light Pollution Reduction? (Choose two)

A. Schools

B. Office buildings

C. Hospitals

D. Museums

E. Dormitories

Q4.19. What percentage of the full-time equivalent (FTE) occupancy should teams determine in order to provide the minimum number of showers and changing rooms in order to comply with SS Credit 3.2, Alternative Transportation, Bicycle Storage and Changing Rooms? (Choose one)

A. 0.15 percent of FTE

B. 0.5 percent of FTE

C. 0.25 percent of FTE

D. 5 percent of FTE

E. 1 percent of FTE

Q4.20. Which of the following are strategies to reduce stormwater runoff? (Choose three)

A. Green roofs

B. Impervious asphalt

C. Pervious pavers

D. Bioswales

Q4.21. Which of the following are LEED concepts that are *most* significantly influenced by site selection prior to the selection of the tenant space? (Choose two)

A. Heat island effect

B. Light pollution reduction

C. Alternative transportation: public transportation access

D. Stormwater management: quality control

E. Brownfield redevelopment

Q4.22. How many basic services are required to comply with SS Credit 2: Development Density and Community Connectivity? (Choose one)

A. 5

B. 4

C. 12

D. 10

E. 6

Q4.23. What strategy is defined in the reference guide that could earn a project one point within the SS Credit 1, Path 2: Stormwater Design, Quality Control if implemented? (Choose two)

A. Increase the perviousness of the site by 50 percent.

B. Improve the runoff quality by removing 90 percent of the total suspended solids.

C. Increase the perviousness of the site by 25 percent.

D. Remove at least 40 percent of the average annual site area's total phosphorus (TP).

E. Improve the runoff quality by removing 80 percent of the TSS.

Q4.24. What roofing material would be best to comply with the SS Credit 1, Path 5 for heat island effect reductions? (Choose two)

A. Gray asphalt with an SRI of 22

B. Aluminum coating with an SRI of 50

C. Red clay tile with an SRI of 36

D. White EPDM with an SRI of 84

E. White cement tile with an SRI of 90

F. Light gravel on built-up roof with an SRI of 37

Q4.25. Using open-grid pavers would contribute to which SS Credit 1: Site Selection path? (Choose two)

A. Path 4: Heat Island Effect, Non-roof

B. Path 2: Stormwater Management: Quantity Control

C. Path 3: Stormwater Management: Quality Control

D. Path 6: Light Pollution Reduction

Q4.26. Which of the following project types are not eligible to pursue SS Credit 2: Development Density and Community Connectivity? (Choose two)

A. A site within an urban neighborhood

B. A greenfield site

C. A previously developed site

D. A graded site

E. One within ½ mile from 5 basic services

Q4.27. Which of the following would qualify as public transportation for the purposes of LEED? (Choose three)

A. Campus bus

B. Light rail

C. Carpooling

D. Ferry service

E. Vanpooling

Q4.28. A project team is pursuing SS Credit 1: Site Selection, Path 5 to reduce the heat island effect. The total flat roof surface is 50,000 square feet, and they plan on installing 15,000 square feet of photovoltaic panels. The mechanical equipment occupies 5,000 square feet. How much of the roof surface has to be covered with a high-SRI material, and what is the minimum SRI of the material in order to comply with the credit? (Choose one)

A. 22,500 SF has to be covered with a roofing material with at least an SRI value of 78

B. 15,000 SF has to be covered with a roofing material with at least an SRI value of 78

C. 25,000 SF has to be covered with a roofing material with at least an SRI value of 78

D. 25,000 SF has to be covered with a roofing material with at least an SRI value of 29

E. 15,000 SF has to be covered with a roofing material with at least an SRI value of 29

Q4.29. A tenant pursuing LEED CI certification has to determine the FTE occupancy for a 312,000-square-foot facility, where they are the only occupant. The new interior space will house 400 full-time employees and 600 part-time employees that work an average of four hours per day. In order to comply with SS Credit 3.2: Bicycle Storage and Changing Rooms, how many showers and bike racks need to be provided? (Choose one)

A. 4 showers and 35 bike racks

B. 3 showers and 21 bike racks

C. 3 showers and 35 bike racks

D. 4 showers and 21 bike racks

Q4.30. What is the maximum allowable distance showers can be located from the building's entrance for a project to comply with SS Credit 3.2: Bicycle Storage and Changing Rooms? (Choose one)

A. 200 feet

B. 300 feet

C. 200 yards

D. 100 feet

E. 100 yards

DESIGN

SS Credit 1:
Site Selection

PURPOSE

To encourage tenants to lease spaces within buildings that implement _____ practices and _____ strategies.

REQUIREMENTS

Select a previously LEED _____ building under one of the following rating systems:

- LEED for New Construction and _____ Renovations

- LEED for _____

- LEED for Core & _____

- LEED for Existing Buildings: _____ & Maintenance

5 pts

RESPONSIBLE PARTY:

OWNER

SS Credit 1:

Site Selection

ANSWER KEY

PURPOSE

To encourage tenants to lease spaces within buildings that implement **best** practices and **green** strategies.

REQUIREMENTS

Select a previously LEED **certified** building under one of the following rating systems:

- LEED for New Construction and **Major** Renovations

- LEED for **Schools**

- LEED for Core & **Shell**

- LEED for Existing Buildings: **Operations** & Maintenance

DESIGN

SS Credit 1: Option 2
Path 1: Brownfield Redevelopment

PURPOSE

To encourage tenants to _____ spaces within buildings that implement best _____ and green _____.

REQUIREMENTS

Option 1:
    Lease a tenant space within a building on a site that was determined to be _____ and was _____.

Option 2:
    Select a building on a site that is _____ as a brownfield by a local, state, or federal _____ agency and where the appropriate _____ was performed.

DOCUMENTATION

Phase _____ report or documentation from _____ agency along with remediation techniques implemented.

REFERENCED STANDARDS

U.S. EPA, _____ of Brownfields

ASTM _____ Phase II Environmental Site Assessment

1 pt

RESPONSIBLE PARTY:
OWNER

SS Credit 1: Option 2
Path 1: Brownfield Redevelopment

ANSWER KEY

PURPOSE

To encourage tenants to **lease** spaces within buildings that implement best **practices** and green **strategies.**

REQUIREMENTS

Option 1:

Lease a tenant space within a building on a site that was determined to be **contaminated** and was **remediated.**

Option 2:

Select a building on a site that is **classified** as a brownfield by a local, state, or federal **government** agency and where the appropriate **remediation** was performed.

DOCUMENTATION

Phase **II** report or documentation from **government** agency along with remediation techniques implemented.

REFERENCED STANDARDS

U.S. EPA, **Definition** of Brownfields

ASTM **E1903-97** Phase II Environmental Site Assessment

DESIGN

**SS Credit 1: Option 2
Path 2, Stormwater Design
Quantity Control**

PURPOSE

To encourage tenants to lease spaces within buildings that implement _____ practices and _____ strategies.

REQUIREMENTS

Option 1: Prior to development, sites with Existing Imperviousness of 50 percent or less

After development discharge _____ and _____ must _____ exceed the rate and quantity _____ to development.

Option 2: Prior to development, site with Existing Imperviousness of more than 50 percent

Decrease the rate and discharge of stormwater runoff by _____ percent of the _____ -year, _____ -hour design storm.

EQUATIONS/DOCUMENTATION

Option 1: Prior to development, sites with Existing Imperviousness of _____ percent or _____

Determine the predevelopment discharge quantity and rate for the site and compare to the postdevelopment quantity and rate.

Option 2: Prior to development, site with Existing Imperviousness of _____ than _____ percent

Determine the predevelopment and postdevelopment discharge quantities and rates for the site and compare to the 1 1/2-year, 24-hour design storm to ensure a _____ percent _____.

The *LEED ID+C Reference Guide* provides runoff coefficients to use for each type of surface for the following calculation:

_____ Area (sf) = Surface Area (sf) x _____ Coefficient

1 pt

**RESPONSIBLE PARTY:
OWNER**

SS Credit 1: Option 2

Path 2, Stormwater Design

Quantity Control

ANSWER KEY

PURPOSE

To encourage tenants to lease spaces within buildings that implement **best** practices and **green** strategies.

REQUIREMENTS

Option 1: Prior to development, sites with Existing Imperviousness of 50 percent or less

    After development discharge **rate** and **quantity** must **not** exceed the rate and quantity **prior** to development.

Option 2: Prior to development, site with Existing Imperviousness of more than 50 percent

    Decrease the rate and discharge of stormwater runoff by **25** percent of the **1 1/2**-year, **24**-hour design storm.

EQUATIONS/DOCUMENTATION

Option 1: Prior to development, sites with Existing Imperviousness of **50** percent or **less**

    Determine the predevelopment discharge quantity and rate for the site and compare to the postdevelopment quantity and rate.

Option 2: Prior to development, site with Existing Imperviousness of **more** than **50** percent

    Determine the predevelopment and postdevelopment discharge quantities and rates for the site and compare to the 1 1/2-year, 24-hour design storm to ensure a **25** percent **decrease**.

The *LEED ID+C Reference Guide* provides runoff coefficients to use for each type of surface for the following calculation:

    Impervious Area (sf) = Surface Area (sf) x **Runoff** Coefficient

DESIGN

SS Credit 1: Option 2
Path 3, Stormwater Design
Quality Control

PURPOSE

To encourage tenants to _____ spaces within buildings that implement best _____ and green _____.

REQUIREMENTS

Lease a tenant space within a building on a site with best management practices (BMPs) that remove at least _____ percent of the average annual site area's total suspended solids _____ and _____ percent of the average annual site area's total phosphorus_____.

BMPs can include the following:

- _____ basins
- Bioswales
- Constructed _____
- Stormwater _____ systems
- _____ filter strips

DOCUMENTATION

Stormwater management plan listing the _____ including structural and nonstructural measures to improve stormwater quality and the contribution of each measure.

REFERENCED STANDARD

U.S. EPA Management Measures for Sources of Non-Point Pollution in Coastal Waters

1 pt

RESPONSIBLE PARTY:
OWNER

SS Credit 1: Option 2
Path 3, Stormwater Design
Quality Control

ANSWER KEY

PURPOSE

To encourage tenants to **lease** spaces within buildings that implement best **practices** and green **strategies**.

REQUIREMENTS

Lease a tenant space within a building on a site with best management practices (BMPs) that remove at least **80** percent of the average annual site area's total suspended solids **(TSS)** and **40** percent of the average annual site area's total phosphorus **(TP)**.
BMPs can include the following:

- **Rentention** basins
- Bioswales
- Constructed **wetlands**
- Stormwater **filtering** systems
- **Vegetated** filter strips

DOCUMENTATION

Stormwater management plan listing the **BMPs** including structural and nonstructural measures to improve stormwater quality and the contribution of each measure.

REFERENCED STANDARD

U.S. EPA Management Measures for Sources of Non-Point Pollution in Coastal Waters

DESIGN

SS Credit 1: Option 2
Path 4, Heat Island Effect
Non-Roof

PURPOSE

To encourage tenants to lease spaces within buildings that implement _____ practices and _____ strategies.

REQUIREMENTS

Option 1: Combine any of the following strategies for 30 percent of the hardscape areas on site:

- _____ within _____ years by the means of trees
- Hardscape surfaces with an SRI of at least _____
- Open-grid _____ system

Option 2:

At least _____ percent of parking is _____ or under _____ by the means of either a green roof or roof surface with an _____ of at least _____.

Option 3:

At least _____ percent of parking has an _____-grid paving system with less than _____ percent impervious.

DOCUMENTATION

Provide a _____ plan depicting nonroof hardscape area square footage and SRI values and/or pervious capabilities for compliant areas.

EXEMPLARY PERFORMANCE

Pursue at least _____ of the compliance options listed above and document under Path _____.

1 pt

RESPONSIBLE PARTY:
OWNER

SS Credit 1: Option 2

Path 4, Heat Island Effect

Non-Roof

ANSWER KEY

PURPOSE

To encourage tenants to lease spaces within buildings that implement **best** practices and **green** strategies.

REQUIREMENTS

Option 1: Combine any of the following strategies for 30 percent of the hardscape areas on site:

- **Shade** within **5** years by the means of trees
- Hardscape surfaces with an SRI of at least **29**
- Open-grid **pavement** system

Option 2:

At least **50** percent of parking is **underground** or under **cover** by the means of either a green roof or roof surface with an **SRI** of at least **29**.

Option 3:

At least **50** percent of parking has an **open**-grid paving system with less than **50** percent impervious.

DOCUMENTATION

Provide a **site** plan depicting nonroof hardscape area square footage and SRI values and/or pervious capabilities for compliant areas.

EXEMPLARY PERFORMANCE

Pursue at least **2** of the compliance options listed above and document under Path **12**.

DESIGN

SS Credit 1: Option 2
Path 5, Heat Island Effect
Roof

PURPOSE

To encourage tenants to _____ spaces within buildings that implement best _____ and green _____.

REQUIREMENTS

Option 1: Lease a tenant space within a building on a site with at least _____ percent of the roof with roofing materials with a minimum _____ value of:

- _____ for _____ -sloped roofs

- __ for _____ -sloped roofs

Option 2: Lease a tenant space within a building on a site with a _____ roof for at least _____ percent of the roof surface

Option 3: Lease a tenant space within a building on a site with a _____ of Option 1 and Option 2

DOCUMENTATION

Calculate total roof area minus any equipment and penetrations, such as skylights.

Provide roof plan with slopes and roofing materials with SRI values and/or vegetated roof areas.

EXEMPLARY PERFORMANCE

Cover _____ percent of the roof with a _____ roof.

REFERENCED STANDARD

_____ International Standards

1 pt

RESPONSIBLE PARTY:
OWNER

SS Credit 1: Option 2
Path 5, Heat Island Effect
Roof

ANSWER KEY

PURPOSE

To encourage tenants to **lease** spaces within buildings that implement best **practices** and green **strategies**.

REQUIREMENTS

Option 1: Lease a tenant space within a building on a site with at least **75** percent of the roof with roofing materials with a minimum **SRI** value of:

- **78** for **low**-sloped roofs
- **29** for **steep**-sloped roofs

Option 2: Lease a tenant space within a building on a site with a **vegetated** roof for at least **50** percent of the roof surface

Option 3: Lease a tenant space within a building on a site with a **combination** of Option 1 and Option 2

DOCUMENTATION

Calculate total roof area minus any equipment and penetrations, such as skylights.

Provide roof plan with slopes and roofing materials with SRI values and/or vegetated roof areas.

EXEMPLARY PERFORMANCE

Cover **100** percent of the roof with a **green** roof.

REFERENCED STANDARD

ASTM International Standards

DESIGN

SS Credit 1: Option 2

Path 6, Light Pollution Reduction

PURPOSE

To encourage tenants to lease spaces within buildings that implement _____ practices and _____ strategies.

REQUIREMENTS

Option 1: Reduced light _____

Locate the tenant space within a building with nonemergency interior light fixtures that are automatically controlled to reduce the light output by _____ percent (or turn off) between the hours of 11:00 pm and _____ am.

Option 2: _____ devices

With an automatically controlled device, ensure _____ envelope openings are shaded to allow a _____ transmittance of no more than _____ percent between the hours of _____ pm and 5:00 am. This is the only compliance option for projects operating _____ hours a day.

STRATEGIES

At all envelope openings:

Option 1: _____ sweep timers, occupancy sensors, programmed master lighting _____ panels

Option 2: Automatic _____ with less than _____ percent _____

DOCUMENTATION

Provide _____ for control or shade locations. Provide building _____ plan or sequence of _____ for automatic shading devices or lighting controls.

Option 2: Provide specifications for product data proving the _____ percent or less transmittance.

1 pt

RESPONSIBLE PARTY:

OWNER

SS Credit 1: Option 2

Path 6, Light Pollution Reduction

ANSWER KEY

PURPOSE

To encourage tenants to lease spaces within buildings that implement **best** practices and **green** strategies.

REQUIREMENTS

Option 1: Reduced light **output**

Locate the tenant space within a building with nonemergency interior light fixtures that are automatically controlled to reduce the light output by **50** percent (or turn off) between the hours of 11:00 pm and **5:00** am.

Option 2: **Shading** devices

With an automatically controlled device, ensure **all** envelope openings are shaded to allow a **light** transmittance of no more than **10** percent between the hours of **11:00** pm and 5:00 am. This is the only compliance option for projects operating **24** hours a day.

STRATEGIES

At all envelope openings:

Option 1: **Automatic** sweep timers, occupancy sensors, programmed master lighting **controls** panels

Option 2: Automatic **shades** with less than **10** percent **transmittance**

DOCUMENTATION

Provide **plans** for control or shade locations. Provide building **operation** plan or sequence of **operation** for automatic shading devices or lighting controls.

Option 2: Provide specifications for product data proving the **10** percent or less transmittance.

DESIGN

SS Credit 1: Option 2
Paths 7 and 8
Water Efficient Landscaping,
50% Reduction and No Potable Water Use/No Irrigation

PURPOSE

To encourage tenants to _____ spaces within buildings that implement best _____ and green _____.

REQUIREMENTS

Path 7: _____ Reduction (2 points)
Lease a tenant space within a building on a site that use _____ percent less _____ water for irrigation by implementing a high efficiency irrigation system and/or uses graywater.
Path 8: No _____ Water Use OR No Irrigation System (2 points)
Use stormwater or graywater to irrigate or plant drought tolerant vegetation that does not require any watering. Temporary irrigation allowed for up to _____ year.

STRATEGIES

Use _____ and adaptive _____ that can survive on natural rainfall.
Use _____ -efficiency irrigation systems.
Use captured _____ or _____ for irrigation.

CALCULATIONS

Calculate _____ demand based on the following factors:
    Square footage of _____ area
    Square footage for each type of _____
    _____ factor, density factor, and _____ factor for each planting type
    _____ rate for climate region
    If irrigation system is to be used, type and controller efficiency

2-4 pts

RESPONSIBLE PARTY:
OWNER

SS Credit 1: Option 2
Paths 7 and 8
Water Efficient Landscaping,
50% Reduction and No Potable Water Use/No Irrigation

ANSWER KEY

PURPOSE

To encourage tenants to **lease** spaces within buildings that implement best **practices** and green **strategies.**

REQUIREMENTS

Path 7: **50%** Reduction (2 points)

Lease a tenant space within a building on a site that use **50** percent less **potable** water for irrigation by implementing a high-efficiency irrigation system and/or uses graywater.

Path 8: No **Potable** Water Use OR No Irrigation System (2 points)

Use stormwater or graywater to irrigate or plant drought tolerant vegetation that does not require any watering. Temporary irrigation allowed for up to **one** year.

STRATEGIES

Use **native** and adaptive **vegetation** that can survive on natural rainfall.

Use **high**-efficiency irrigation systems.

Use captured **stormwater** or **graywater** for irrigation.

CALCULATIONS

Calculate water demand based on the following factors:

    Square footage of **landscaped** area

    Square footage for each type of **vegetation**

    Species factor, density factor, and **microclimate** factor for each planting type

    Evapotranspiration rate for climate region

    If irrigation system is to be used, type and controller efficiency

DESIGN

SS Credit 1: Option 2
Path 9
Innovative Wastewater Technologies

PURPOSE

To encourage tenants to lease spaces within buildings that implement _____ practices and _____ strategies.

REQUIREMENTS

Option 1:
  Select a base building that reduces potable water used for sewage conveyance by _____ percent by implementing water-conserving _____ or using nonpotable water sources, such as graywater and captured stormwater.

Option 2:
  Choose a tenant space within a building that treats _____ percent of wastewater on-site to _____ standards to be infiltrated or used-on-site to recharge the aquifer.

REFERENCED STANDARDS

The _____ Policy Act (EPAct) of _____

The Energy _____ Act (EPAct) of 2005

The _____ Plumbing Code or _____ Plumbing Code

2 pts

RESPONSIBLE PARTY:
OWNER

SS Credit 1: Option 2

Path 9

Innovative Wastewater Technologies

ANSWER KEY

PURPOSE

To encourage tenants to lease spaces within buildings that implement **best** practices and **green** strategies.

REQUIREMENTS

Option 1:

Select a base building that reduces potable water used for sewage conveyance by **50** percent by implementing water-conserving **fixtures** or using nonpotable water sources, such as graywater and captured stormwater.

Option 2:

Choose a tenant space within a building that treats **100** percent of wastewater on-site to **tertiary** standards to be infiltrated or used-on-site to recharge the aquifer.

REFERENCED STANDARDS

The **Energy** Policy Act (EPAct) of **1992**

The Energy **Policy** Act (EPAct) of 2005

The **Uniform** Plumbing Code or **International** Plumbing Code

DESIGN

SS Credit 1: Option 2
Path 10
30% Water Use Reduction

PURPOSE

To encourage tenants to _____ spaces within buildings that implement best _____ and green _____.

REQUIREMENTS

Lease a tenant space in a base building that reduces potable water use by at least _____ percent for the _____ building with a _____ for future occupants to _____.

Install low-flow and low-flush fixtures to reduce water consumption.

Calculate _____ consumption using occupancy uses for each fixture type based on _____ values.

Calculate _____ case consumption using _____ uses for each fixture type based on _____ fixture values.

Assume a _____:1 gender ratio for calculations.

EXEMPLARY PERFORMANCE

Lease a tenant space in a base building that reduces potable water use by at least _____ percent for the _____ building with a _____ for future occupants to _____.

REFERENCED STANDARDS

The Energy Policy Act (_____) of 1992
The Energy _____ Act (EPAct) of 2005
The Uniform Plumbing _____ or International _____ Code

1 pt

RESPONSIBLE PARTY:
OWNER

SS Credit 1: Option 2
Path 10
30% Water Use Reduction

ANSWER KEY

PURPOSE

To encourage tenants to **lease** spaces within buildings that implement best **practices** and green **strategies.**

REQUIREMENTS

Lease a tenant space in a base building that reduces potable water use by at least **30** percent for the **entire** building with a **plan** for future occupants to **comply.**

Install low-flow and low-flush fixtures to reduce water consumption.

Calculate **baseline** consumption using occupancy uses for each fixture type based on **EPAct** values.

Calculate **design** case consumption using **occupancy** uses for each fixture type based on **actual** fixture values.

Assume a **1**:1 gender ratio for calculations.

EXEMPLARY PERFORMANCE

Lease a tenant space in a base building that reduces potable water use by at least **40** percent for the **entire** building with a **plan** for future occupants to **comply.**

REFERENCED STANDARDS

The Energy Policy Act **(EPAct)** of 1992

The Energy **Policy** Act (EPAct) of 2005

The Uniform Plumbing **Code** or International **Plumbing** Code

DESIGN

SS Credit 1: Option 2
Path 11
On-Site Renewable Energy

PURPOSE

To encourage tenants to _____ spaces within buildings that implement _____ practices and _____ strategies.

REQUIREMENTS

Lease a tenant space within a building that has an eligible type of a _____ energy system on-site.

The system must supply at least _____ percent (1 point) or _____ percent (2 points) of the building's _____ energy use.

STRATEGIES

Eligible Systems include:

   Solar thermal, _____ systems, wind, wave, some biomass systems, _____ (heating and electric) power, and low-impact hydropower systems

Ineligible systems include:

   Architectural features, _____ solar strategies, _____ strategies, and geo-exchange systems using a _____ -source heat pump

EXEMPLARY PERFORMANCE

The system must supply at least _____ percent of the building's _____ energy use.

REFERENCED STANDARD

ANSI/ASHRAE/IESNA Standard _____

1–2 pts

RESPONSIBLE PARTY:
OWNER

	SS Credit 1: Option 2
	Path 11
	On-Site Renewable Energy

ANSWER KEY

PURPOSE
To encourage **tenants** to lease spaces within buildings that implement **best** practices and **green** strategies.

REQUIREMENTS
Lease a tenant space within a building that has an eligible type of a **renewable** energy system on-site.
The system must supply at least **2.5** percent (1 point) or **5** percent (2 points) of the building's **total** energy use.

STRATEGIES
Eligible Systems include:
Solar thermal, **photovoltaic** systems, wind, wave, some biomass systems, **geothermal** (heating and electric) power, and low-impact hydropower systems
Ineligible systems include:
Architectural features, **passive** solar strategies, **daylighting** strategies, and geo-exchange systems using a **ground**-source heat pump

EXEMPLARY PERFORMANCE
The system must supply at least **10** percent of the building's **total** energy use.

REFERENCED STANDARD
ANSI/ASHRAE/IESNA Standard **90.1**

DESIGN

SS Credit 1: Option 2

Path 12

Other Quantifiable Environmental Performance

PURPOSE

To encourage _____ to lease spaces within buildings that implement best _____ and green _____.

REQUIREMENTS

Document _____ performance from Paths 1– _____ or achievement of the requirements of _____ LEED rating system credit.

1 pt

RESPONSIBLE PARTY:

OWNER

SS Credit 1: Option 2
Path 12
Other Quantifiable Environmental Performance

ANSWER KEY

PURPOSE

To encourage **tenants** to lease spaces within buildings that implement best **practices** and green **strategies**.

REQUIREMENTS

Document **exemplary** performance from Paths 1**–11** or achievement of the requirements of **another** LEED rating system credit.

DESIGN

SS Credit 2:
Development Density and Community Connectivity

PURPOSE

To encourage the development or redevelopment of high-density areas with _____ infrastructure, in order to preserve greenfield sites, _____, and natural resources.

REQUIREMENTS

Option 1: _____ Density

　　Select a site within an existing neighborhood with a density of _____ square feet per acre, including the project seeking certification

Option 2: Community _____

- Select a site within a half of mile of a residential area with an average density of 10 units per acre

AND

- Develop a site within a half mile of _____ basic services with pedestrian access

　　_____ out 10 basic services must be existing

　　_____ can be counted twice

　　_____ use projects can include only _____ service within LEED boundary

6 pts

RESPONSIBLE PARTY:

OWNER

SS Credit 2:
Development Density and Community Connectivity

ANSWER KEY

PURPOSE

To encourage the development or redevelopment of high-density areas with **existing** infrastructure, in order to preserve greenfield sites, **habitats**, and natural resources.

REQUIREMENTS

Option 1: **Development** Density

> Select a site within an existing neighborhood with a density of 60,000 square feet per acre, including the project seeking certification

Option 2: Community **Connectivity**

- Develop a site within a half of mile of a residential area with an average density of 10 units per acre

AND

- Develop a site within a half mile of 10 basic services with pedestrian access

    8 out 10 basic services must be existing

    Restaurants can be counted twice

    Mixed use projects can include only **one** service within LEED boundary

DESIGN

SS Credit 3.1:
Alternative Transportation,
Public Transportation Access

PURPOSE

Reduce the negative environmental impacts from _____ use including _____ and _____ development.

REQUIREMENTS

Option 1: Rail

    Site is within a _____ mile of a commuter rail, light rail, rail, tram, or ferry station.

Option 2: Bus

    Site is within a _____ of a mile of _____ or more bus (public or campus) lines.

EQUATIONS/DOCUMENTATION

    Site _____ plan with walking distances labeled.

EXEMPLARY PERFORMANCE

Option 1: Rail

    Site is within a half mile of at least two existing commuter rail, light rail, rail, tram, or ferry station.

    200 _____ rides per day at these stops.

Option 2: Bus

    Site is within a quarter of a mile of _____ or more more bus (public or campus) lines.

    _____ transit rides per day at these stops.

Both options:

    Conduct a _____ Transportation _____ Plan in order to reduce a quantifiable amount of car use by implementing alternative options.

6 pts

RESPONSIBLE PARTY:
OWNER

SS Credit 3.1: Alternative Transportation, Public Transportation Access

ANSWER KEY

PURPOSE

Reduce the negative environmental impacts from **car** use including **pollution** and **land** development.

REQUIREMENTS

Option 1: Rail

    Site is within a **half** mile of a commuter rail, light rail, rail, tram, or ferry station.

Option 2: Bus

    Site is within a **quarter** of a mile of **two** or more bus (public or campus) lines.

EQUATIONS/DOCUMENTATION

    Site **vicinity** plan with walking distances labeled.

EXEMPLARY PERFORMANCE

Option 1: Rail

    Site is within a half mile of at least two existing commuter rail, light rail, rail, tram, or ferry station.

    200 **transit** rides per day at these stops.

Option 2: Bus

    Site is within a quarter of a mile of four or more bus (public or campus) lines.

    200 transit rides per day at these stops.

Both options:

    Conduct a **Comprehensive** Transportation **Management** Plan in order to reduce a quantifiable amount of car use by implementing alternative options.

DESIGN	**SS Credit 3.2:**
	Alternative Transportation,
	Bicycle Storage and Changing Rooms

PURPOSE

Reduce the negative environmental impacts from _____ use, including _____ and _____ development.

REQUIREMENTS

Select a site that provides at least _____ percent of tenant occupants with bike storage and _____ percent of FTE with showers within _____ yards of building entrance.

    Showers and changing rooms that are located in a nearby _____ club must be _____ of charge to tenants. _____ year contract agreement required.

EQUATIONS/DOCUMENTATION

All Options:

    Create a plan with bike storage and shower and changing room locations in relationship to _____ entrance.

    Calculate tenant, _____, and transient occupants. FTE = Total staff occupant hours / _____

EXEMPLARY PERFORMANCE

    Conduct a _____ Transportation _____ Plan in order to reduce a quantifiable amount of car use by implementing alternative options.

2 pts

RESPONSIBLE PARTY:
ARCHITECT

SS Credit 3.2:
Alternative Transportation,
Bicycle Storage and Changing Rooms

ANSWER KEY

PURPOSE

Reduce the negative environmental impacts from **car** use, including **pollution** and **land** development.

REQUIREMENTS

Select a site that provides at least **5** percent of tenant occupants with bike storage and **0.5** percent of FTE with showers within **200** yards of building entrance.

Showers and changing rooms that are located in a nearby **health** club must be **free** of charge to tenants. **2** year contract agreement required.

EQUATIONS/DOCUMENTATION

All Options:

Create a plan with bike storage and shower and changing room locations in relationship to **building** entrance.

Calculate tenant, **FTE**, and transient occupants. FTE = Total staff occupant hours / **8**

EXEMPLARY PERFORMANCE

Conduct a **Comprehensive** Transportation **Management** Plan in order to reduce a quantifiable amount of car use by implementing alternative options.

DESIGN

SS Credit 3.3:
Alternative Transportation, Parking Capacity

PURPOSE

Reduce the negative environmental impacts from _____ use, including _____ and _____ development.

REQUIREMENTS

Case 1: Tenants that lease less than 75% of total building area

    Option 1: Parking spaces allotted to _____ occupants must not exceed _____ zoning ordinance. Project to also provide _____ parking for ride share vehicles for carpools/vanpools equal to _____ percent of tenant occupants.

    Option 2: Do not provide any parking for tenant occupants.

Case 2: Tenants that lease 75% or more of total building area

    Option 1: Parking capacity must not exceed local _____ ordinance. Project to also provide preferred parking for ride share vehicles for carpools/vanpools equal to _____ percent of _____ occupants.

    Option 2: Do not provide any _____ parking for rehabilitation projects. Provide preferred parking for _____ percent of the building _____.

EXEMPLARY PERFORMANCE

    Conduct a _____ Transportation _____ Plan in order to reduce a quantifiable amount of car use by implementing alternative options.

2 pts

RESPONSIBLE PARTY:
ARCHITECT

SS Credit 3.3

Alternative Transportation, Parking Capacity

ANSWER KEY

PURPOSE

Reduce the negative environmental impacts from car use, including pollution and land development.

REQUIREMENTS

Case 1: Tenants that lease less than 75% of total building area

    Option 1: Parking spaces allotted to **tenant** occupants must not exceed **local** zoning ordinance. Project to also provide **preferred** parking for ride share vehicles for carpools/vanpools equal to **5** percent of tenant occupants.

    Option 2: Do not provide any parking for tenant occupants.

Case 2: Tenants that lease 75% or more of total building area

    Option 1: Parking capacity must not exceed local **zoning** ordinance. Project to also provide preferred parking for ride share vehicles for carpools/vanpools equal to **5** percent of **building** occupants.

    Option 2: Do not provide any **new** parking for rehabilitation projects. Provide preferred parking for **5** percent of the building **occupants.**

EXEMPLARY PERFORMANCE

    Conduct a Comprehensive Transportation Management Plan in order to reduce a quantifiable amount of car use by implementing alternative options.

CHAPTER 5
WATER EFFICIENCY

THIS CHAPTER FOCUSES ON THE STRATEGIES and technologies described within the Water Efficiency (WE) category of the Leadership in Energy and Environmental Design (LEED®) for Commercial Interiors™ (CI) rating system, including methods to reduce the consumption of water, our most precious resource, which is often taken for granted. As the demand for water continues to increase and supplies continue to decrease, it is challenging for municipalities to keep up. The U.S. Geological Survey estimates that buildings account for 12 percent of total water use in the United States. Potable water that is delivered to buildings and homes is first pulled from local bodies of water, treated, and then delivered. This water is typically used for toilets, urinals, sinks (Figure 5.1), showers, drinking, irrigation, and for equipment uses, such as mechanical systems, dishwashers, and washing machines. Once the wastewater leaves the building or home, it is treated and then delivered back to the body of water. When the influx supersedes the capacity of the wastewater treatment facilities, overflow will result. This overflow can pollute and contaminate nearby water bodies, the sources of potable water, therefore causing the need for more treatment facilities to be built. Therefore, it is critical to understand how to reduce the amount of water we consume, to reduce the burden on the entire cycle, especially as we are threatened with shortages in the near future.

Green interior design teams have the opportunity to specify efficient fixtures, equipment, and appliances that require less water. Within the LEED CI rating system, water efficiency for green interior spaces is addressed in terms of indoor water use. The previous chapter also addressed water efficiency strategies but bear in mind SS Credit 1 addresses the base building whereas this category specifically addresses the interior leased space.

 Remember to pick a new color for flashcards created for the WE category topics.

Figure 5.1 Implementing commercial bathroom lavatories with sensors help to reduce water consumption and could help to achieve WE Prerequisite 1 of the LEED CI rating system.
Photo courtesy of Stephen Martorana, LEED AP BD+C

The indoor water use reduction strategy is referred to in both the prerequisite and credit of the category, as shown in Table 5.1. Notice both can be submitted after construction documents are completed for a design-side review. Do not worry about remembering the exact credit name and number, as the exam will list them together, but it is important to know the amount of points available and both refer to achievements in reducing the amount of potable water demanded for indoor fixtures and use specifically within the tenant space.

Table 5.1 The Water Efficiency Category

D/C	Prereq/Credit	Title	Points
D	Prereq 1	Water Use Reduction – 20% Reduction	R
D	Credit 1	Water Use Reduction	6 to 11

INDOOR WATER USE

Indoor water use typically includes the water used for water closets, urinals, lavatories, and showers. Break room or kitchen sinks are also included in the calculations for indoor water use. For the purposes of the exam, it is important to understand and remember the differences between a flush fixture and a flow fixture and how the consumption of the fixture is measured. Flush fixtures, such as toilets and urinals (Figure 5.2), are measured in **gallons per flush** (gpf). Flow fixtures, such as sink faucets, showerheads, and aerators, are measured in **gallons per minute** (gpm).

Indoor Water Use in Relation to LEED Compliance

When approaching the strategies to reduce water for a project seeking LEED certification, it is necessary for the project teams to calculate a **baseline** for water usage to compare to the amount the project is intended to require. The WE prerequisite and some of the credits utilizes the Energy Policy Act of 1992 (EPAct 1992) for flow and flush rates associated with conventional and efficient fixtures (see Table 5.2). Project teams should also reference the Energy Policy Act of 2005 (EPAct 2005), as it became U.S. law in August 2005.

Once the fixture water consumption is determined, project teams need to account for the occupant usage to calculate how much water is required for the building. Remember from the previous chapter, the full-time equivalent (FTE) occupancy is an estimation of actual tenant space occupancy in terms of hours occupied per day and is used to determine the number of occupants for the building that will use the fixtures. FTE is calculated by dividing the total number of occupant hours spent in the building (each full-time employee is assumed to be in the building for 8 hours) divided by eight. Therefore, full-time employees have a value of one. Part-time employees must also be considered in the calculations, if they work four hours a day, they have a value of 0.5. For example, if a building has 100 occupants, 50 of whom work full time and 50 occupants of whom work part time, the FTE for the project is 75.

Keeping within the lines of the basic concept to use less water, efficient indoor water strategies help to change the typically traditional, wasteful behavior of

Water efficiency helps to reduce energy and, therefore, costs by reducing the amount of water that must be treated, heated, cooled, and distributed.

Create flashcards to remember flow and flush fixture types and how they are measured.

Create a flashcard to remember the Energy Policy Act of 1992 (EPAct 1992) and EPAct 2005 as the standards for the WE prerequisite and indoor water use credit.

Notice that FTE is the acronym for **full-time equivalent** and not *full-time employee*. Make sure you account for part-time occupants as well!

A low-flow water closet uses 30 percent less water than a conventional water closet.

Figure 5.2 Waterless urinals help to reduce the indoor water consumption.
Photo courtesy of SmithGroup, Inc.

occupants. Project teams should conduct life-cycle cost assessments for determining the best solution for their projects. For example, waterless urinals might cost less to install, as they do not require water, but their maintenance costs might be higher than those of a conventional fixture. Most of these strategies will not be noticeable, but they will substantially reduce water consumption.

Create a flashcard to remember baseline versus design case: the amount of water a conventional project would use as compared to the design case.

Table 5.2 Water Consumption Assumptions According to EPAct 1992

Fixture Type	Gallons per Flush (gpf)
Conventional water closet (for baseline calculations)	1.6
Low-flow water closet	1.1
Ultra-low-flow water closet	0.8
Composting toilet	0.0
Conventional urinal (for baseline calculations)	1.0
Waterless urinal	0.0

WE Prerequisite 1: Water Use Reduction. The *ID+C Reference Guide* defines potable water use as an important component that green buildings should address and, therefore, requires a reduction in consumption as a minimal performance feature. This importance is characterized by the means of a prerequisite within the WE category. As a result, LEED-certified interior projects must demand at least 20 percent or less indoor water compared to conventionally designed interior spaces. The strategy to achieve this prerequisite involves implementing water efficient flush and flow fixtures (Figure 5.3) within the tenant space. As previously stated, project teams would need to calculate a baseline water consumption using the EPAct values based on the occupancy with a 1:1 male-to-female gender ratio. They will then need to input the values specific to the fixtures specified for the specific project to then compare the different demanded consumptions to prove a 20 percent reduction. These

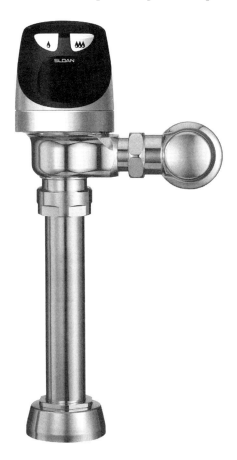

Figure 5.3 Using high-efficiency faucets and high-efficiency toilet (HET) fixtures and flushometers, which use 1.28 gpf or less, helps to achieve the water reduction prerequisite of 20 percent.
Photo courtesy of Sloan Valve Company

INDOOR WATER USE

Figure 5.4 Water-efficient dishwashers or washing machines can help a team to pursue exemplary performance for WE Credit 2.
Photo courtesy of Gettys

calculations are completed on the credit form on LEED-Online. Remember the built-in functionality that would determine compliance is based on the data input.

WE Credit 1: Water Use Reduction. In order to comply with this credit, project teams will need to exceed the minimum threshold required by the prerequisite. Project teams can earn up to 11 points for achieving 30 percent (6 points), 35 percent (8 points), and 40 percent (11 points) indoor water use reduction. The can pursue an exemplary performance opportunity to reducing water demand by 45 percent to earn another point. To help project teams reach the 45 percent reduction minimum, dishwashers, laundry machines, and other water-consuming fixtures can be included in the calculations (Figure 5.4). Including signage to help educate the tenant space occupants can also help the team to earn an Innovation in Design (ID) credit (Figure 5.5).

 TIP If a tenant space and/or project scope does not include any plumbing fixtures, the project is exempt from complying with the prerequisite.

 Create a flashcard to remember the two opportunities that address indoor water use.

Figure 5.5 The Armstrong World Industries Corporate Headquarters in Lancaster, Pennsylvania, implemented signage to help educate the occupants, a strategy that helped the project earn Platinum certification.
Photo courtesy of Armstrong Ceiling and Wall Systems

QUIZ TIME!

Q5.1. Which of the following uses are best described and suitable for nonpotable water? (Choose two)

    A. Drinking water

    B. Irrigation

    C. Clothes washing

    D. Process water

    E. Dishwashing

    F. Showers

Q5.2. Which of the following products are not examples of a flow fixture? (Choose two)

    A. Lavatory faucets

    B. Toilets

    C. Sprinkler heads

    D. Aerators

    E. Showerheads

    F. Urinals

Q5.3. What is the minimum percentage of reduced water consumption required in order to comply with WE Prerequisite 1: Water Use Reduction? (Choose one)

    A. 10 percent

    B. 15 percent

    C. 20 percent

    D. 25 percent

    E. 30 percent

Q5.4. Which of the following is addressed for compliance with WE Credit 1: Water Use Reduction? (Choose three)

    A. Showerheads

    B. Sprinkler heads

    C. Hose bibs

    D. Janitor sinks

    E. Kitchen faucets

Q5.5. How many possible points are available to a LEED CI project team for reducing potable water usage within the tenant space? (Choose one)

    A. 6 points

    B. 8 points

    C. 11 points

    D. 12 points

    E. 13 points

DESIGN

WE Prerequisite 1:
Water Use Reduction
20% Reduction

WE Credit 1:
Water Use Reduction
30% to 40% Reduction

PURPOSE

To ensure a reduction on the _____ of the municipal water supply and _____ systems by _____ water _____ within buildings.

REQUIREMENTS

Install low-flow and low-flush fixtures to _____ water consumption.

Calculate _____ consumption using occupancy uses for each fixture type based on _____ values.

Calculate _____ case consumption using occupancy uses for each fixture type based on _____ fixture values.

Assume a _____ gender ratio for calculations.

POINT DISTRIBUTION

20%	30%	35%	40%	45%
• Required	• 6 points	• 8 points	• 11 points	• 1 EP point

REFERENCED STANDARDS

The _____ Policy Act (EPAct) of 1992

The _____ Policy Act (EPAct) of 2005

The _____ Plumbing Code or International _____ Code

2–4 pts

RESPONSIBLE PARTY:
PLUMBING ENGINEER

WE Prerequisite 1:
Water Use Reduction
20% Reduction

WE Credit 1:
Water Use Reduction
30% to 40% Reduction

ANSWER KEY

PURPOSE

To ensure a reduction on the **burden** of the municipal water supply and **wastewater** systems by **increasing** water **efficiency** within buildings.

REQUIREMENTS

Install low-flow and low-flush fixtures to **reduce** water consumption.

Calculate **baseline** consumption using occupancy uses for each fixture type based on **EPAct** values.

Calculate **design** case consumption using occupancy uses for each fixture type based on **actual** fixture values.

Assume a **1:1** gender ratio for calculations.

POINT DISTRIBUTION

20%	30%	35%	40%	45%
• Required	• 6 points	• 8 points	• 11 points	• 1 EP point

REFERENCED STANDARDS

The **Energy** Policy Act (EPAct) of 1992

The **Energy** Policy Act (EPAct) of 2005

The **Uniform** Plumbing Code or International **Plumbing** Code

CHAPTER 6
ENERGY AND ATMOSPHERE

THIS CHAPTER FOCUSES ON THE STRATEGIES AND TECHNOLOGIES to address energy use and consumption as described in the Energy & Atmosphere (EA) category of the Leadership in Energy and Environmental Design (LEED®) for Commercial Interiors™ (CI) rating system (Figure 6.1). By now, we all understand the environmental impacts of using fossil fuels to generate electricity. Each step of the electricity production process harms the environment and ecosystem in one way or another. For example, the burning of coal releases harmful pollutants and greenhouse gases that contribute to global warming and climate change, reducing air quality on a global scale.

Remember from Chapter 2 that conventionally designed and built facilities account for 39 percent primary energy use, 72 percent electricity consumption, and 38 percent carbon dioxide (CO_2) emissions, according to the U.S. Green Building Council (USGBC) website.[1] Therefore, the LEED rating systems put the most emphasis on the EA category by offering the largest opportunity to earn points, in an attempt to reduce the electrical consumption and corresponding CO_2 emissions of certified buildings. Project teams are encouraged to focus on the following four components in order to address the goals and intentions of the EA category to help reduce greenhouse gas emissions:

 TIP Burning coal releases the following harmful pollutants into the atmosphere: carbon dioxide, sulfur dioxide, nitrogen oxide, and mercury.

 TIP It's time to pick a different color for flashcards created for EA category topics.

Figure 6.1 The BP America project in DC, 1101 NY Ave, by FOX Architects incorporates daylight harvesting to optimize its energy consumption to help earn the project a Platinum certification under the LEED for Commercial Interiors rating system. *Photo courtesy of Ron Blunt Photography*

1. Energy efficiency
2. Tracking energy consumption
3. Managing refrigerants
4. Renewable energy

> **TIP** Create a flashcard to remember the four components of the Energy & Atmosphere category.

Each of these strategies is referred to in the prerequisites and credits of the category, as shown in Table 6.1. Remember from the previous chapters to note when the prerequisite or credit is eligible to be submitted for certification review, at the end of design or construction. So as you can see, only the commissioning prerequisite and credit and Credit 4 are not eligible for a design-side review. Also note that Credit 1.3 Optimize Energy Performance – HVAC offers the most point opportunity in the category. Do not worry about remembering the exact credit name and number, but it is important to know which are credits versus prerequisites.

Table 6.1 The Energy & Atmosphere Category

D/C	Prereq/Credit	Title	Points
C	Prereq 1	Fundamental Commissioning of Building Energy Systems	R
D	Prereq 2	Minimum Energy Performance	R
D	Prereq 3	Fundamental Refrigerant Management	R
D	Credit 1.1	Optimize Energy Performance – Lighting Power	1 to 5
D	Credit 1.2	Optimize Energy Performance – Lighting Controls	1 to 3
D	Credit 1.3	Optimize Energy Performance – HVAC	5 to 10
D	Credit 1.4	Optimize Energy Performance – Equipment and Appliances	1 to 4
C	Credit 2	Enhanced Commissioning	5
D	Credit 3	Measurement and Verification	2 to 5
C	Credit 4	Green Power	5

ENERGY EFFICIENCY

The integrative design process is critical within all the LEED categories, but it is essential within the EA category, especially when determining the right components to deliver the correct holistic solution to reduce energy demand (Figure 6.2). Typically, a tenant is not granted with the opportunity to work with a developer to design a base building, but instead leases space within an existing building. Regardless, there are certain components to note when selecting a base building as energy performance, demands, and requirements are affected, including:

- Site conditions, such as materials that contribute to heat island reduction, can reduce energy demand, as equipment will not need to compensate for heat gain from surrounding and adjacent areas.

- Building orientation can affect the amount of energy needed for artificial heating, cooling, and lighting needs by taking advantage of free energy by the

Figure 6.2 The Pennsylvania Department of Conservation and Natural Resources' Penn Nursery project incorporates radiant heat flooring to optimize its energy consumption.
Photo courtesy of Moshier Studio

means of passive design strategies, such as daylighting, natural ventilation, and implementing a trombe wall (Figure 6.3).

- How much water needs to be heated or cooled? If building system equipment and fixtures require less water, less energy is required. If all of the building equipment is sized appropriately and works efficiently, then less energy is demanded (Figures 6.4 and 6.5).

- Shifting loads to off-peak periods can help to reduce demand. Using thermal energy storage to take advantage of the temperature fluctuations associated with the day time versus the night, allows project teams to refuse heat at

 TIP Minimize solar gain in the summer and maximize it in the winter with the help of passive design strategies! Passive designs capitalize on the four natural thermal processes: radiation, conduction, absorption, and convection.

Figure 6.3 The Wetland Discovery Point building at Utah State University utilizes a trombe wall to capture heat from the sun, as well as radiant heat flooring to increase the energy efficiencies of the project.
Photo courtesy of Gary Neuenswander, Utah Agricultural Experiment Station

Figure 6.4 High-efficiency boilers can help to achieve energy performance and savings goals. *Photo courtesy of SmithGroup, Inc.*

> **TIP** Remember, SRI is the acronym for **solar reflectance index** and is synonymous with *albedo*. Do you remember the scale used for SRI? Is it better to have a higher or lower score?

night to provide cooling during the day in the summer and capture heat during the day to use at night in the winter (Figure 6.6).

- Roof design can impact how much energy is required for heating and cooling by implementing a green roof or a roof with a high solar reflectance index (SRI) value.

- Building envelope thermal performance, including window selections, can reduce mechanical system sizing and energy demands by ensuring a thermal break between the interior and exterior environments (Figure 6.7). Insulating

Figure 6.5 Installing high-efficiency chillers can help to achieve energy performance and savings goals.
Photo courtesy of SmithGroup, Inc.

a building envelope can help to reduce the size of heating, ventilation, and air conditioning (HVAC) systems, and thus help to use less energy (Figure 6.8).

- Light fixture types and the lamps/bulbs they require can reduce energy use by providing more light per square foot but require fewer kilowatts per hour and, therefore, optimize **lighting power density,** or the amount of lighting power installed per unit area.

Figure 6.6 Ice ball thermal energy storage helps to provide cooling during the day from ice generated at night to reduce energy demands.
Photo courtesy of Cryogel

Figure 6.7 Selecting insulated concrete forms (ICFs) as a building envelope material can increase the performance of a building.
Photo courtesy of Moshier Studio

Figure 6.8 Spray foam insulation helps to seal any leaks of conditioned air to the exterior environment, thus optimizing energy use.
Photo courtesy of BioBased Technologies

The preceding bullet points describe the importance and need for project team members to find a base building that works in a cohesive fashion optimizing the performance of the building and its site, ultimately reducing the amount of energy required for operations. Tenant improvement project teams are encouraged to take advantage of energy modeling and simulation software to study and evaluate how their specific project will function. Teams that use Building Information Modeling (BIM) software have an advantage in determining synergistic opportunities for their projects and in adding efficiencies. Both of these design phase studies also contribute to whole-building life-cycle cost assessments to determine trade-offs between up-front costs and long-term savings.

 TIP The key to energy modeling and simulation is whole-building evaluation, not individual component assessments. How do all of the systems work together?

Energy Efficiency in Relation to LEED Compliance

Optimizing energy performance is presented as a prerequisite and a credit suite in order to require a minimum level of efficiency to address the energy demand of green buildings, helping to save energy and, thus, reducing greenhouse gas emissions and saving money.

EA Prerequisite 2: Minimum Energy Performance. The LEED CI rating system requires tenant spaces to perform to a minimum energy standard. Project teams have the opportunity to choose from a performance-based compliance path or a prescriptive approach. The performance-based option requires an energy model to depict the assumed performance of the building and related energy consumption. Similarly to the reference standard Energy Policy Act of 1992 (EPAct 1992, with the subsequent rulings of EPAct 2005), as discussed in the previous chapter, which is used to create a **baseline** for comparison to the **design case**, a baseline is needed for this performance-based compliance path to evaluate energy use reduction percentages of a project. Therefore, LEED references American Society of Heating, Refrigerating, and Air-Conditioning Engineers (ASHRAE) Standard 90.1-2007, Appendix G, to determine a baseline energy performance requirement for buildings seeking LEED CI certification. Project energy models are required to demonstrate compliance with the mandatory provisions of the standard and a minimum of 10 percent reduction in lighting power density (LPD) below the standard. The scope of work within the tenant's responsibility will also need to include ENERGY STAR® equipment for 50 percent (by rated power) of the equipment installed. Therefore, base building equipment is exempt from compliance.

 Create a flashcard to remember ASHRAE 90.1-2007, Appendix G, as the baseline standard for energy performance. However, if the local code is more stringent, such as California's Title 24-2005, it can be used instead of the ASHRAE standard.

EA Credit 1.1: Optimize Energy Performance – Lighting Power. Taking the requirements of the prerequisite a step further, the LPD will need to be reduced by at least 15 percent to earn this credit (Figure 6.9). Point earning potential increases for every 5 percent reduction. Teams can pursue an Exemplary Performance credit under the Innovation in Design (ID) category should the reduction surpass 40 percent.

 TIP Any time you see ASHRAE 90.1, think ENERGY!

 TIP Remember, only performance-based compliance paths require energy modeling; prescriptive-based compliance path options do not.

"This credit compares the installed interior lighting power with the interior lighting power allowance" within the tenant space.[2] First, the lighting designer or engineer will need to calculate the connected lighting power using ASHRAE 90.1 standard methodology by multiplying the quantity of luminare type by the luminare wattage (by type). Project teams will then need to calculate the lighting power allowance by using the space-by-space method or the building area method.

Figure 6.9 The lighting strategies at the Rockefeller Brothers Fund interior space in New York designed by Illumination Arts and FXFOWLE Architects include decorative wallwashers between the windows and task lights at each desk. This lighting strategy was able to reduce the lighting power density (LPD) by more than 25 percent and earned the project three LEED points.
Photo courtesy of FXFOWLE Architects and Eric Laignel

The space-by-space method takes each type of space within the tenant's scope of work to determine the LPD. The space types are listed in the reference standard which includes a corresponding LPD allowance in watts per square foot. The project's allowance is generated by "multiplying the allowed lighting power density of each space area type by the gross lighted floor area of that building type."[3]

The building area method, is a more holistic approach as it takes the building area types determined by the reference standard and establishes an allowed LPD in watts per square foot. The allowance is generated by "multiplying the allowed lighting power density of each building area type by the gross lighted floor area of that building type."[4] This approach seems to be a bit more complicated, but it is beneficial if the tenant space has more display or decorative lighting.

After the interior lighting power is determined, the lighting power reduction is calculated by subtracting the installed/connected interior lighting power from the interior lighting power allowance.

EA Credit 1.2: Optimize Energy Performance – Lighting Controls. Project teams have 3 opportunities to meet the intentions of this credit to earn up to 3 points. For one point, daylight controls will need to be installed "in all regularly occupied daylit spaces within 15 feet of windows and skylights."[5] The second opportunity requires daylight responsive controls to be installed for at least 50 percent of the connected lighting load. A point can also be earned should the project implement occupancy sensors for at least 75 percent of the connected lighting load (Figure 6.10). The team can pursue an extra point for exemplary performance should the design include daylight responsive controls for at least 75 percent of the connected lighting load or should the design include occupancy sensors for at least 95 percent of the connected lighting load (both scenarios require a 25 percent increase from the credit requirements).

ENERGY EFFICIENCY 117

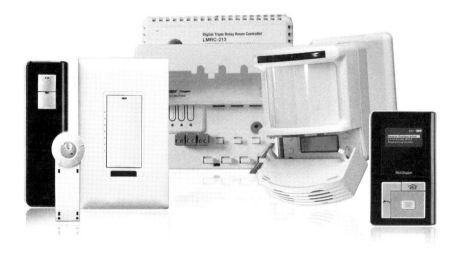

Figure 6.10 Implementing lighting controls, such as sensors, can help to satisfy the requirements of EA Credit 1.2.
Photo courtesy of Wattstopper

EA Credit 1.3: Optimize Energy Performance – HVAC. The largest point opportunity of the category, the LEED CI rating system provides two compliance options for teams to choose from to earn up to 10 points. Option 1 allows teams to address equipment efficiency for five points and zoning and controls for another five points (Figure 6.11). The equipment efficiency option requires to the heating, ventilation, and air-conditioning (HVAC) systems to comply with New Building's *Advanced Buildings™ Core Performance™ Guide*'s mechanical equipment efficiency and variable speed control standards. The appropriate zoning and

 Be sure to refer to the study worksheets at the end of the chapter to review the equations for installed lighting power, interior lighting power allowance, lighting power reduction, and lighting power density percentage reduction.

Figure 6.11 Implementing chilled beams can help to reduce the energy demand of tenant spaces.
Photo courtesy of Swegon

Special occupancy areas can include break rooms, kitchens, and conference rooms.

controls option requires every solar exposure to have its own separate zone, all interior spaces to be separately zoned, and private offices and special occupancy areas to have active controls that can sense use and modulate the HVAC system to accommodate demand. This is more of a prescriptive approach and, therefore, no calculations are required.

Option 2 also allows two opportunities to earn up to 10 points and has the only exemplary performance point potential. To comply and earn 5 points, teams need to document at least a 15 percent reduction energy cost as compared to ASHRAE 90.1's **energy cost budget** for regulated energy. Should they be able to prove at least a 30 percent reduction, they can pursue another 5 points. Exemplary performance is awarded should the energy model demonstrate the tenant space is 33 percent more efficient than the reference standard's energy cost budget. This option requires an energy model to prove compliance. Energy modeling typically involves seven steps:

 Regulated energy includes space heating, space cooling, and associated fans and pumps within the tenant space for the purposes of Option 2 compliance.

1. Select a modeler.
2. Determine the portion of the building to model (as it will typically include more than the tenant space).
3. Select a modeling approach (depending if the base building is already modeled).
4. Acquire building information such as as-built drawings, operational schedules, HVAC system and zones, and lighting systems.
5. Model the design case.
6. Model the baseline case using the referenced standards mandatory provisions and prescriptive requirements.
   a. Model an alternative baseline case (only if the base building is more efficient than the reference standard).

 Create a flashcard to remember the seven steps of energy modeling.

7. Calculate the energy reduction.

EA Credit 1.4: Optimize Energy Performance – Equipment and Appliances. Taking the requirements of the prerequisite a step further, teams will need to install at least 70 percent (by rated power) ENERGY STAR–eligible equipment and appliances to earn 1 point. They can earn up to 4 points for installing 90 percent qualifying equipment. An exemplary performance opportunity exists under the ID category should the tenant scope of work include 97 percent ENERGY STAR–eligible equipment.

 EA Credit 1.4 applies to appliances, office equipment, electronics, and commercial food equipment. The credit does not address HVAC, lighting, and building envelope products.

 Create a flashcard to remember the prerequisite and 4 credits for energy efficiency strategies within the LEED CI rating system.

QUIZ TIME!

Q6.1. What is the minimum lighting power density reduction percentage reduction required in order to comply with EA Credit 1.1: Optimize Energy Performance – Lighting Power? (Choose one)

A. 2 percent

B. 4 percent

C. 15 percent

D. 10 percent

E. 12 percent

ENERGY EFFICIENCY

Q6.2. What percentage of ENERGY STAR equipment is required to be implemented to comply with the EA energy performance prerequisite? (Choose one)

  A. 12 percent
  B. 10 percent
  C. 5 percent
  D. 50 percent

Q6.3. What is the minimum lighting power density (LPD) reduction percentage required in order to meet the EA energy performance prerequisite? (Choose one)

  A. 2 percent
  B. 4 percent
  C. 5 percent
  D. 10 percent
  E. 12 percent

Q6.4. What is the reference standard to refer to in order to comply with the *performance*-based path of EA Credit 1.3: Optimize Energy Performance – HVAC? (Choose one)

  A. Whole-building simulation
  B. *Advanced Buildings™ Core Performance™ Guide*
  C. *ASHRAE Advanced Energy Design Guide*
  D. ENERGY STAR
  E. ASHRAE 90.1-2007

Q6.5. What is the reference standard to refer to in order to comply with the *prescriptive*-based path of EA Credit 1.3: Optimize Energy Performance – HVAC? (Choose one)

  A. Whole-building simulation
  B. Advanced Buildings™ Core Performance™ Guide
  C. ASHRAE Advanced Energy Design Guide
  D. ENERGY STAR
  E. ASHRAE 90.1-2007

Q6.6. In order to comply with EA Credit 1.2: Optimize Energy Performance – Lighting Controls, which of the following strategies should a team employ? (Choose two)

  A. Continuously dimmed daylight responsive lighting
  B. Step dimming
  C. Occupancy sensors for 50 percent of the connected lighting load
  D. Daylight controls for 25 percent of the lighting load
  E. Daylight responsive controls within 25 feet of windows

TRACKING ENERGY CONSUMPTION

The LEED CI rating system details a number of means for project teams to address the monitoring and tracking of energy consumption. Projects seeking LEED certification are required not only to be designed and constructed to perform to a minimal energy performance but also to implement procedures to ensure the performance, such as **commissioning** (Cx). Commissioning a building ensures that the equipment and systems are performing as they were intended, to maintain consistent and minimal energy demands. Just as with energy performance, the EA category addresses Cx in terms of a prerequisite and a credit opportunity.

The EA category also includes another credit for the project teams to address the tracking of energy consumption, the Measurement and Verification credit. By providing meters and other technology to track the consumption of energy, operation and maintenance staff members are able to detect inefficiencies, such as excessive demand and, thus, take corrective action. Metering utilities also ensures the ongoing accountability of energy usage (Figure 6.12).

Tracking Energy Consumption in Relation to LEED Compliance

EA Prerequisite 1: Fundamental Commissioning of Building Energy Systems. The first prerequisite of the EA category requires the tenant's scope of work to be commissioned by a **commissioning authority** (CxA). The CxA reports directly to the owner and represents the owner, responsible for coordinating with the design team and then the contractor during construction. Note, because of the role a CxA plays, he or she can be an employee of the developer/owner but may not be the engineer of record (see Study Tip). He or she must have experience commissioning at least two other projects to be qualified as the CxA on a project seeking LEED certification.

The commissioning process begins *early* in the design process in which the CxA works with the owner to establish the **owner's project requirements** (OPR).

TIP A commissioning agent (CxA) cannot be the engineer of record for the project. Think about it—who would double-check the engineer's design if he or she were the same person? For projects smaller than 50,000 square feet and not pursuing EA Credit 2, a qualified employee of one of the design team firms is eligible as long as he or she is not the same professional responsible for engineering services.

Figure 6.12 Metering utilities to monitor consumption helps building owners and facility managers ensure that the building is functioning properly. *Photo courtesy of Gary Neuenswander, Utah Agricultural Experiment Station*

The OPR includes the environmental goals of the project and is issued to the design team to develop a **basis of design** (BOD) for the major building systems, such as lighting; domestic hot water; heating, ventilation, air conditioning, and refrigeration (HVAC&R), and any renewable energy generated on site (that is part of the tenant's scope). The BOD includes a description of these systems, the applicable indoor environmental quality measures, and any assumptions as related to the design. The document should also include any codes, ordinances, and regulations applicable to the project. The Cx process continues with the development of the Cx plan and incorporating the Cx requirements into the construction documents (CDs). The CxA works diligently during construction to ensure that the building system equipment is installed and calibrated and performs appropriately and efficiently to avoid construction defects and equipment malfunctions by the means of **systems performance testing.** Finally, the CxA is responsible for completing a summary Cx report in order to comply with this prerequisite.

EA Credit 2: Enhanced Commissioning. Table 6.2 summarizes all of the Cx activities for both the prerequisite and the credit. As the table indicates, the reference guide requires five extra tasks to be completed by the CxA in order to achieve this credit. These additional tasks begin prior to the completion of the CDs, as the CxA is required to review the design drawings and specifications to avoid design flaws and ensure that the environmental goals are included, such as water and energy use reductions, and they are in line with the OPR and BOD. The second task requires the CxA to review the construction submittals for the applicable systems to be commissioned. After functional testing of the systems is completed, but before the building is occupied, the CxA develops operation and maintenance manuals and helps to educate the facility management teams on the operation and maintenance strategies specific to the building (Figure 6.13). Finally, within 10 months after occupancy, the CxA returns to the site to ensure that the building systems are working accordingly and address any needed adjustments.

> **TIP** Remember, prerequisites are absolutely required, do not contribute to earning points, and ensure that certified projects meet minimum performance criteria.

> **TIP** Commissioning has an average payback of 4.8 years and typically costs about $1 per square foot, according to a study conducted by Lawrence Berkley National Laboratory.[6]

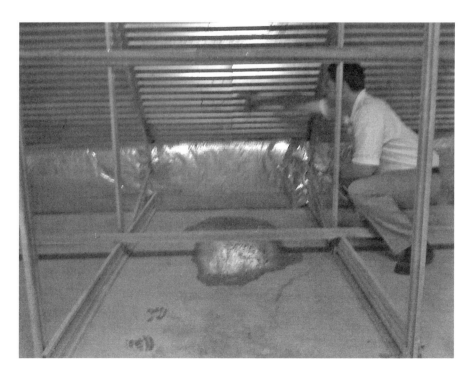

Figure 6.13 Inspecting building systems and educating the operations and maintenance staff on how the systems are intended to operate helps to ensure that a building performs the way it was designed.
Photo courtesy of ENERActive Solutions

Create a flashcard to remember the benefits of utilizing a CxA:
1. Minimize or eliminate design flaws.
2. Avoid construction defects.
3. Avoid equipment malfunctions.
4. Ensure that preventative maintenance is implemented during operations.

Table 6.2 Fundamental Commissioning versus Enhanced Commissioning

Phases	Task	Prerequisite 1: Fundamental Cx	Credit 2: Enhanced Cx
Predesign, Design	1. Designate CxA	Owner or project team	
	2. Document OPR and then develop BOD	Owner or CxA; design team	
	3. Review OPR and BOD	CxA	
	4. Develop and implement a Cx plan	Project team or CxA	
	5. Incorporate Cx requirements into CDs	Project team or CxA	
	6. Conduct Cx design review prior to midconstruction documents	—	CxA
Construction	7. Review contractor submittals applicable to systems being commissioned	—	CxA
	8. Verify installation and performance of commissioned systems	CxA	
	9. Develop systems manual for commissioned systems	—	Project team or CxA
	10. Verify that requirements for training are completed	—	Project team or CxA
	11. Complete a summary commissioning report	CxA	
Occupancy	12. Review building operation within 8 to 10 months after substantial completion	—	CxA

Be sure to remember the 5 tasks associated with EA Credit 2 separate from the prerequisite.

EA Credit 3: Measurement and Verification. Project teams seeking to achieve this credit will need to determine the appropriate compliance path based on the size of the tenant space. If a tenant space is less than 75 percent of the total building area, Case 1 should be pursued. In this instance, teams will need to complete at least one of the following for 2 to 5 points:

- Install submetering equipment for all energy sources (2 points).
- Energy costs should be paid by the tenant (3 points).

Should the tenant space occupy 75 percent or more of the total building area, the following 3 requirements need to be fulfilled to be awarded the 5 points under Case 2:

- Install continuous metering equipment.
- Develop and implement a measure and verification (M&V) plan in accordance with the International Performance Measure & Verification Protocol (IPMVP), Volume III.
- Provide an approach for corrective action should the anticipated energy savings not be achieved during operation.

Create two flashcards to remember the occupancy determination for EA Credit 3 and the requirements of the two compliance options.

Create a flashcard to remember the prerequisite and 2 credits LEED utilizes to address the monitoring and tracking of energy consumption.

QUIZ TIME!

Q6.7. Which of the following is not required to be included in the BOD? (Choose one)

A. Indoor environmental quality criteria

B. Applicable codes

C. Tenant design guidelines

D. Energy-related system descriptions

E. Design assumptions

Q6.8. When should the OPR be prepared? (Choose one)

A. Schematic Design

B. Construction Documents

C. Design Development

D. Beginning of Construction

E. Substantial Completion

Q6.9. Which of the following are appropriate statements with regard to commissioning in the context of green building? (Choose two)

A. The CxA should be a primary member of the design team who is directly responsible for the project design or construction management.

B. The CxA should be separate and independent from those individuals who are directly responsible for project design or construction management (preferably from a separate firm).

C. The CxA is indirectly responsible for verifying the performance of building systems and equipment prior to installation, calibration, and operation.

D. The CxA is responsible for verifying the performance of building systems and equipment after installation.

Q6.10. A tenant space occupies 82 percent of the total building area. Which of the following are required to be uploaded to LEED-Online for certification review to be in compliance with EA Credit 3: Measurement and Verification? (Choose two)

A. Tenant design guidelines

B. Plan with metering locations

C. Agreement with local utility provider

D. Meter type, manufacturer, and model number

E. Measurement and verification plan

F. Lease agreement for tenant

Q6.11. What is the referenced standard for EA Credit 3: Measurement and Verification? (Choose one)

A. ASHRAE 90.1–2007
B. *ASHRAE Advanced Energy Design Guide*
C. *IPMVP*, Volume III
D. *Advanced Buildings™ Core Performance™ Guide*

MANAGING REFRIGERANTS

Besides the commissioning and minimum energy performance prerequisites, the LEED CI rating system also requires buildings to manage **refrigerants** appropriately. Refrigerants enable the transfer of thermal energy and are, therefore, a critical component of air-conditioning and refrigeration equipment for their ability to reject heat at high temperatures. Although they are cost-effective, refrigerants have environmental trade-offs, as they contribute to ozone depletion and global warming. Therefore, project teams need to be mindful of the ozone-depleting potential (ODP) and global warming potential (GWP) of each refrigerant to determine the impact of the trade-offs, as an environmentally perfect refrigerant does not yet exist.

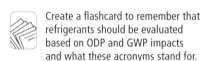
Create a flashcard to remember that refrigerants should be evaluated based on ODP and GWP impacts and what these acronyms stand for.

Managing Refrigerants in Relation to LEED Compliance

EA Prerequisite 3: Fundamental Refrigerant Management. To comply with this prerequisite, teams should refer to the Montreal Protocol when determining which refrigerants to use for their projects. The Montreal Protocol bans **chlorofluorocarbons** (CFCs), as they have the biggest impact on ozone depletion, and therefore their use is not allowed when pursuing certification. When the tenant scope of work includes the installation of a HVAC system, teams will need to comply with this prerequisite. Should the scope of work not include modifying an existing system or utilizes natural ventilation strategies, the project is exempt from complying.

QUIZ TIME!

Q6.12. Which of the following refrigerants is not allowed to be used for HVAC systems when pursuing LEED CI certification? (Choose one)

A. CFCs
B. HCFCs
C. HFCs
D. Halons

Q6.13. Which of the following is characterized as regulated energy? (Choose three)

- A. Office equipment
- B. Exhaust fans
- C. Elevators
- D. Space heating
- E. Space cooling

Q6.14. When addressing refrigerants for a project and to comply with EA Prerequisite 3: Fundamental Refrigerant Management, which of the following should be considered? (Choose two)

- A. New HVAC systems
- B. Base building air-conditioning systems
- C. Boilers for heating systems
- D. Reuse of existing HVAC&R systems

Q6.15. Which of the following is not required to comply with Prerequisite 1: Fundamental Commissioning of Building Energy Systems? (Choose three)

- A. Develop the BOD
- B. Develop a systems manual
- C. Develop the owner's project requirements
- D. Inclusion of commissioning requirements into the construction documents
- E. Verify building operation training
- F. CxA to review submittals during construction

RENEWABLE ENERGY

In keeping with the goals previously discussed, implementing renewable energy technologies in a green building project can reduce the need to produce and consume coal, nuclear power, and oil and natural gases for energy, thus reducing pollutants and emissions, as well as increasing air quality. Remember on-site renewable energy was presented in the Sustainable Sites Category in Chapter 4, as it is assumed to be an owner-driven decision and implementation and not the tenant's. LEED CI provides the tenant with one strategy to address renewable energy and reduce the use of fossil fuels for projects seeking LEED certification.

EA Credit 4: Green Power. In order to comply with this credit, project owners need to purchase energy from a Green-e certified provider or **renewable energy certificates (RECs)** from a Green-e eligible source. The renewable energy associated with this credit is generated at a different location than the project site seeking LEED certification (Figure 6.14). Project teams need to determine the energy demand of the project by the estimations of an energy model or use a default electrical consumption value (see Study Tip). Tenants need to then purchase 50 percent of the project's estimated energy demand for at least two years based on quantity, not the cost of energy, to comply (as with EA Credit 1.3). If they purchase 100 percent or more, the project could seek an exemplary performance point.

Create a flashcard to remember the meaning of the acronym REC.

TIP Teams can either purchase 50 percent of the energy model expectation or use the default rate of 8 kilowatt-hours per square foot per year. Should they use the default value of 16 kilowatt-hours per square foot per year, the team can pursue an exemplary performance point.

Create a flashcard to remember the one credit associated with renewable energy for LEED CI project teams to pursue.

Figure 6.14 Cedar Creek Wind Farm in Colorado helps to produce clean, renewable energy. *Photo courtesy of Brian Stanback, Renewable Choice Energy*

QUIZ TIME!

Q6.16. ASHRAE Standard 90.1-2007 is primarily concerned with which of the following? (Choose one)

  A. Lighting design
  B. Ventilation effectiveness
  C. Energy consumption
  D. Ozone depletion

Q6.17. Which of the following are categorized as regulated energy? (Choose three)

  A. Computers
  B. Building envelope
  C. Refrigeration
  D. Service hot water
  E. Lighting

Q6.18. What is the first step to determine lighting power density percentage reduction? (Choose one)

  A. Subtract installed interior lighting power from interior lighting power allowance.
  B. Calculate interior lighting power allowance with the space-by-space method.
  C. Determine the connected/installed lighting power.
  D. Calculate the interior lighting power allowance with the building area method.

Q6.19. What is the primary intent of EA Credit 4: Green Power? (Choose one)

   A. To comply with the Montreal Protocol

   B. To encourage more solar farms in the United States and avoid carbon trading

   C. To encourage the development and use of renewable clean energy that is connected to the utility grid

   D. To minimize production of greenhouse gases by generating on-site renewable energy

Q6.20. Which of the following tasks are not required to comply with EA Prerequisite 2: Enhanced Commissioning? (Choose two)

   A. Submittal review

   B. Operation and maintenance system manuals

   C. Commissioning plan

   D. Basis of design

   E. Functional testing procedures

Q6.21. What is the intent of EA Prerequisite 3: Fundamental Refrigerant Management? (Choose one)

   A. To establish a minimum level of energy efficiency to reduce environmental and economic impacts associated with excessive use in tenant spaces

   B. To reduce stratospheric ozone depletion

   C. To provide for the accountability and optimization of tenant utility consumption over time

   D. To encourage the development and use of grid-source, renewable energy systems on a net zero pollution basis

Q6.22. Which of the following are eligible for exemplary performance opportunities? (Choose two)

   A. EA Prerequisite 2: Minimum Energy Performance

   B. EA Credit 3: Measurement and Verification

   C. EA Credit 4: Green Power

   D. EA Credit 1.1: Optimize Energy Performance – Lighting Power

   E. EA Credit 2: Enhanced Commissioning

Q6.23. Which of the following can be used in place of ASHRAE 90.1–2007 when pursuing EA Credit 1.1: Optimize Energy Performance – Lighting Power? (Choose one)

   A. Space-by-space method

   B. California's Title 24-2005

   C. IPMVP, Volume III

   D. *Advanced Buildings™ Core Performance™ Guide*

   E. Whole building method

Q6.24. Which of the following strategies can be used to reduce lighting power density? (Choose two)

    A. Low-efficacy sources

    B. High-efficacy sources

    C. High internal reflectances

    D. Low internal reflectances

Q6.25. Which of the following is required to comply with Case 1 for projects occupying less than 75 percent of the total building area under EA Credit 3: Measurement and Verification? (Choose two)

    A. Tenant design guidelines

    B. Plan with metering locations

    C. Lease agreement with building owner acknowledging the tenant's responsibility to pay utility cost as it is not included in the base rent

    D. Submetering equipment

    E. Measurement and verification plan

CONSTRUCTION	EA Prerequisite 1: Fundamental Commissioning of Building Energy Systems
	EA Credit 2: Enhanced Commissioning

PURPOSE

Prerequisite:

To ensure the _____-related systems of the building are _____, _____, and are performing accordingly and in line with the owner's _____ requirements, _____ of design, and construction documents. Commissioning benefits include a reduction in _____ use, _____ costs, and contractor _____, an improvement in building documentation and occupant _____, and the verification of system _____.

Credit:

To start the commissioning process _____ in the design phases and complete additional tasks after the systems' performance has been _____.

REQUIREMENTS

Prerequisite – Complete the following _____ tasks:

1. Engage a commissioning agent (CxA)
    - Must have commissioned at least _____ previous projects.
    - CxA must report directly to the owner.
    - Projects smaller than _____ gross square feet and not pursuing EA Credit 2, the CxA may be part of the design or construction teams but not the _____ of _____.
2. Record the owner's project _____ (OPR) and then the _____ of design (BOD) with the design team.
3. The commissioning activities must be included in the construction specifications.
4. A _____ plan must be created and implemented.
5. The installation and performance of the building's energy-related systems must be verified. These systems include:
    - HVAC & R systems and controls
    - Lighting and _____ controls
    - Domestic _____ water systems
    - On-site _____ energy systems
6. The _____ report must be completed.

Credit – In addition to above listed tasks for prerequisite, complete the following _____ tasks:

1. CxA must review the design documents prior to mid-_____ document phase and ensure comments are addressed in _____ submission.
2. CxA must review the construction _____ for the _____-related systems to be commissioned to ensure alignment with _____ and _____.
3. Create a _____ manual.
4. Ensure building operations and maintenance staff have received required _____.
5. The CxA must return to the site within 8 to _____ months after substantial completion to review the building's performance and establish a plan to resolve any outstanding commissioning-related issues.

STRATEGIES

Credit:

When selecting a CxA they must be an independent _____ party although can be an employee of the _____.

They cannot be an employee of the design firm but can be a consultant.

The CxA must not be an employee or a consult to the _____ or _____ manager.

5 pts for credit

RESPONSIBLE PARTY:
COMMISSIONING AGENT

EA Prerequisite 1:
Fundamental Commissioning of Building Energy Systems

EA Credit 2:
Enhanced Commissioning

ANSWER KEY

PURPOSE

Prerequisite:

To ensure the **energy**-related systems of the building are **installed**, **calibrated**, and are performing accordingly and in line with the owner's **project** requirements, **basis** of design, and construction documents.

Commissioning benefits include a reduction in **energy** use, **operating** costs, and contractor **callbacks**, an improvement in building documentation and occupant **productivity**, and the verification of system **performance**.

Credit:

To start the commissioning process **early** in the design phases and complete additional tasks after the systems' performance has been verified.

REQUIREMENTS

Prerequisite – Complete the following **6** tasks:

1. Engage a commissioning agent (CxA)
    - Must have commissioned at least two previous projects.
    - CxA must report directly to the owner.
    - Projects smaller than **50,000** gross square feet and not pursuing EA Credit 2, the CxA may be part of the design or construction teams but not the **engineer** of **record**.
2. Record the owner's project **requirements** (OPR) and then the **basis** of design (BOD) with the design team.
3. The commissioning activities must be included in the construction specifications.
4. A commissioning plan must be created and implemented.
5. The installation and performance of the building's energy-related systems must be verified. These systems include:
    - HVAC & R systems and controls
    - Lighting and **daylighting** controls
    - Domestic **hot** water systems
    - On-site **renewable** energy systems
6. The **commissioning** report must be completed.

Credit – In addition to above listed tasks for prerequisite, complete the following **5** tasks:

1. CxA must review the design documents prior to mid-**construction** document phase and ensure comments are addressed in **subsequent** submission.
2. CxA must review the construction submittals for the **energy**-related systems to be commissioned to ensure alignment with OPR and BOD.
3. Create a **systems** manual.
4. Ensure building operations and maintenance staff have received required **training**.
5. The CxA must return to the site within 8 to **10** months after substantial completion to review the building's performance and establish a plan to resolve any outstanding commissioning-related issues.

STRATEGIES

Credit:

- When selecting a CxA they must be an independent **third** party although can be an employee of the **owner**.
    - They cannot be an employee of the design firm but can be a consultant.
    - The CxA must not be an employee or a consult to the **contractor** or **construction** manager.

DESIGN

**EA Prerequisite 2:
Minimum Energy Performance**

PURPOSE

To ensure a _____ level of performance to increase efficiency and to reduce the environmental and _____ impacts from _____ energy demand.

REQUIREMENTS

Within the tenant's scope of work, design to comply with the mandatory provisions of ASHRAE _____ and achieve the _____ - or _____ - based requirements of the referenced standard.

Within tenant space (or scope) reduce lighting _____ density by at least _____ percent below _____ 90.1.

Install at least _____ percent (by _____ power) ENERGY _____ – eligible equipment within tenant's space.

REFERENCED STANDARDS

ANSI/ASHRAE/IESNA Standard _____

_____ STAR® _____ Products

Required

RESPONSIBLE PARTY:
MECHANICAL ENGINEER

EA Prerequisite 2:
Minimum Energy Performance

ANSWER KEY

PURPOSE

To ensure a **minimum** level of performance to increase efficiency and to reduce the environmental and **economic** impacts from **excessive** energy demand.

REQUIREMENTS

Within the tenant's scope of work, design to comply with the mandatory provisions of ASHRAE **90.1** and achieve the **prescriptive** or **performance**-based requirements of the referenced standard.

Within tenant space (or scope) reduce lighting **power** density by at least **10** percent below **ASHRAE** 90.1.

Install at least **50** percent (by **rated** power) ENERGY **STAR**–eligible equipment within tenant's space.

REFERENCED STANDARDS

ANSI/ASHRAE/IESNA Standard 90.1

ENERGY STAR® **Qualified** Products

	EA Prerequisite 3:
DESIGN	**Fundamental Refrigerant Management**

PURPOSE

To reduce the depletion of the _____.

REQUIREMENTS

The use of _____ - based refrigerants is prohibited in HVAC&R systems within the _____ scope of work.

Refrigerant examples include CFCs, _____, HFCs, and natural refrigerants such as _____ and carbon _____.

Required

RESPONSIBLE PARTY:

MECHANICAL ENGINEER

EA Prerequisite 3:
Fundamental Refrigerant Management

ANSWER KEY

PURPOSE

To reduce the depletion of the **stratosphere**.

REQUIREMENTS

The use of **CFC**-based refrigerants is prohibited in HVAC&R systems within the **tenant's** scope of work.

Refrigerant examples include CFCs, **HCFCs**, HFCs, and natural refrigerants such as **ammonia** and carbon **dioxide**.

DESIGN

EA Credit 1.1:
Optimize Energy Performance, Lighting Power

PURPOSE
To achieve more efficiencies beyond the referenced _____ to further reduce the environmental and economic impacts from excessive energy use.

REQUIREMENTS
Within tenant space (or scope) reduce lighting _____ density by at least _____ percent below _____ 90.1 calculated using wither the Building _____ Method or _____ -by-Space Method.

POINT DISTRIBUTION

15%	20%	25%	30%	35%	40%
• 1 point	• 2 points	• 3 points	• 4 points	• 5 points	• 1 EP Point

EQUATIONS

Equation 1. Installed Interior Lighting Power

Installed Interior Lighting Power (watts) = Quantity by type of luminaire X Luminaire wattage by luminaire type (watts)

Equation 2. Interior Lighting Power (Building Area Method)

Installed Interior Lighting Power Allowance (watts) = Gross lighting floor area (sf) X Building Area Type Lighting Power Density (watts/sf)

Equation 3. Interior Lighting Power (Space-by-Space Method)

Installed Interior Lighting Power Allowance (watts) = Gross lighting floor area (sf) X Space Area Type Lighting Power Density (watts/sf)

Equation 4. Lighting Power Reduction

Lighting Power Reduction (watts) = Interior Lighting Power Allowance (watts) − Installed Interior Lighting Power (watts)

Equation 5. Lighting Power Density Percentage Reduction

$$\text{Percentage Reduction (\%)} = \frac{\text{Lighting Power Reduction (watts)}}{\text{Interior Lighting Power Allowance (watts)}}$$

EXEMPLARY PERFORMANCE
Within tenant space (or scope) reduce lighting power density by at least _____ percent below ASHRAE _____.

REFERENCED STANDARD
ANSI/ASHRAE/IESNA Standard _____

1–5 pts

RESPONSIBLE PARTY:
LIGHTING DESIGNER

EA Credit 1.1: Optimize Energy Performance, Lighting Power

ANSWER KEY

PURPOSE
To achieve more efficiencies beyond the reference **standard** to further reduce the environmental and economic impacts from excessive energy use.

REQUIREMENTS
Within tenant space (or scope) reduce lighting **power** density by at least **15** percent below **ASHRAE** 90.1 calculated using either the Building **Area** Method or **Space**-by-Space Method.

POINT DISTRIBUTION

- 15% • 1 point
- 20% • 2 points
- 25% • 3 points
- 30% • 4 points
- 35% • 5 points
- 40% • 1 EP Point

EQUATIONS

Equation 1. Installed Interior Lighting Power

Installed Interior Lighting Power (watts) = Quantity by type of luminaire X Luminaire wattage by luminaire type (watts)

Equation 2. Interior Lighting Power (Building Area Method)

Installed Interior Lighting Power Allowance (watts) = Gross lighting floor area (sf) X Building Area Type Lighting Power Density (watts/sf)

Equation 3. Interior Lighting Power (Space-by-Space Method)

Installed Interior Lighting Power Allowance (watts) = Gross lighting floor area (sf) X Space Area Type Lighting Power Density (watts/sf)

Equation 4. Lighting Power Reduction

Lighting Power Reduction (watts) = Interior Lighting Power Allowance (watts) − Installed Interior Lighting Power (watts)

Equation 5. Lighting Power Density Percentage Reduction

$$\text{Percentage Reduction (\%)} = \frac{\text{Lighting Power Reduction (watts)}}{\text{Interior Lighting Power Allowance (watts)}}$$

EXEMPLARY PERFORMANCE
Within tenant space (or scope) reduce lighting power density by at least **40** percent below ASHRAE **90.1.**

REFERENCED STANDARD
ANSI/ASHRAE/IESNA Standard **90.1**

	EA Credit 1.2:
DESIGN	**Optimize Energy Performance, Lighting Controls**

PURPOSE

To achieve more efficiencies beyond the _____ standard to further reduce the _____ and _____ impacts from excessive energy use.

REQUIREMENTS

Within tenant space (or scope) implement at least _____ of the following lighting _____ system strategies (_____ point each):

- Daylight _____ controls within all regularly occupied daylight areas within _____ feet of windows or under skylights

- _____ responsive controls for at least _____ percent of the connected lighting load

- _____ sensors for _____ percent of the _____ lighting load

EXEMPLARY PERFORMANCE

Within tenant space (or scope) implement at least _____ of the following lighting _____ system strategies:

- _____ responsive controls for at least _____ percent of the connected lighting load

- _____ sensors for _____ percent of the _____ lighting load

1–3 pts   RESPONSIBLE PARTY: **LIGHTING DESIGNER**

EA Credit 1.2:
Optimize Energy Performance, Lighting Controls

ANSWER KEY

PURPOSE

To achieve more efficiencies beyond the **referenced** standard to further reduce the **environmental** and **economic** impacts from excessive energy use.

REQUIREMENTS

Within tenant space (or scope) implement at least **one** of the following lighting **control** system strategies (**one** point each):

- Daylight **responsive** controls within all regularly occupied daylight areas within **15** feet of windows or under skylights

- **Daylight** responsive controls for at least **50** percent of the connected lighting load

- **Occupancy** sensors for **75** percent of the **connected** lighting load

EXEMPLARY PERFORMANCE

Within tenant space (or scope) implement at least **one** of the following lighting **control** system strategies:

- **Daylight** responsive controls for at least **75** percent of the connected lighting load

- **Occupancy** sensors for **95** percent of the **connected** lighting load

	EA Credit 1.3:
DESIGN	**Optimize Energy Performance, HVAC**

PURPOSE

To achieve more efficiencies beyond the referenced _____ to further reduce the environmental and economic impacts from _____ energy use.

REQUIREMENTS

Option 1. Prescriptive-Based Approach. Implement at least one of the following (5 points each):

- HVAC system that complies with the New Buildings Institute's *Advanced Buildings™ Core Performance Guide* for mechanical _____ efficiency and _____ speed control.

- Install a separate _____ zone for every _____ exposure.

    _____ zone interior spaces.

    Private offices and other special occupancy areas must sense _____ to respond to space _____ .

Option 2. Performance-Based Approach

Design HVAC systems to reduce energy cost by _____ percent as compared with _____ 90.1 (5 points).

Design HVAC systems to reduce energy cost by _____ percent as compared with ASHRAE _____ (10 points).

EXEMPLARY PERFORMANCE

Option 2.

Design HVAC systems to reduce energy cost by at least _____ percent as compared with the referenced _____ .

REFERENCED STANDARD

ANSI/ASHRAE/IESNA Standard _____

_____ Buildings _____ Advanced Buildings™ _____ Performance _____

5–10 pts

RESPONSIBLE PARTY:
MECHANICAL ENGINEER

EA Credit 1.3:
Optimize Energy Performance, HVAC

ANSWER KEY

PURPOSE

To achieve more efficiencies beyond the referenced **standard** to further reduce the environmental and economic impacts from **excessive** energy use.

REQUIREMENTS

Option 1. Prescriptive-Based Approach. Implement at least one of the following (5 points each):

- HVAC system that complies with the New Buildings Institute's *Advanced Core Performance Guide* for mechanical **equipment** efficiency and **variable** speed control.

- Install a separate **control** zone for every **solar** exposure.

    Separately zone interior spaces.

    Private offices and other special occupancy areas must sense **occupancy** to respond to space **demand.**

Option 2. Performance-Based Approach.

    Design HVAC systems to reduce energy cost by **15** percent as compared with **ASHRAE** 90.1 (5 points).

    Design HVAC systems to reduce energy cost by **30** percent as compared with ASHRAE **90.1** (10 points).

EXEMPLARY PERFORMANCE

Option 2.

    Design HVAC systems to reduce energy cost by at least **33** percent as compared with the referenced **standard.**

REFERENCED STANDARD

ANSI/ASHRAE/IESNA Standard **90.1**

New Buildings **Institute's** Advanced **Core** Performance **Guide**

DESIGN

EA Credit 1.4:
Optimize Energy Performance, Equipment and Appliances

PURPOSE

To achieve more efficiencies beyond the _____ standard to further reduce the _____ and _____ impacts from excessive energy use.

REQUIREMENTS

Install at least _____ percent (by rated _____) _____ STAR–eligible equipment within the tenant's space.

POINT DISTRIBUTION

- 70% • 1 point
- 77% • 2 points
- 84% • 3 points
- 90% • 4 points
- 97% • 1 EP Point

EXEMPLARY PERFORMANCE

Install at least _____ percent (by rated _____) _____ STAR–eligible equipment within the tenant's space.

1–4 pts

RESPONSIBLE PARTY: OWNER

EA Credit 1.4:
Optimize Energy Performance, Equipment and Appliances

ANSWER KEY

PURPOSE

To achieve more efficiencies beyond the **referenced** standard to further reduce the **environmental** and **economic** impacts from excessive energy use.

REQUIREMENTS

Install at least **70** percent (by rated **power**) **ENERGY** STAR–eligible equipment within the tenant's space.

POINT DISTRIBUTION

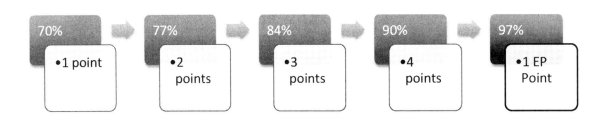

| 70% | 77% | 84% | 90% | 97% |
| •1 point | •2 points | •3 points | •4 points | •1 EP Point |

EXEMPLARY PERFORMANCE

Install at least **97** percent (by rated **power**) **ENERGY** STAR–eligible equipment within the tenant's space.

CONSTRUCTION

EA Credit 3:
Measurement and Verification

PURPOSE

To provide the opportunity to track the tenant's _____ and _____ consumption over time.

REQUIREMENTS

Case 1: Tenants that lease _____ than _____ of total building area (complete at least _____ of the following):

        Install _____ equipment (2 points).

        Tenant _____ for energy costs (not included in base rent) (5 points).

Case 2: Tenants that lease 75% or more of total building area (5 points):

        Install _____ metering equipment.

        Create and implement a _____ & verification (M&V) plan.

        Develop a plan for _____ action if expected energy savings are not achieved.

REFERENCED STANDARD

International Performance _____ & _____ Protocol (IPMVP), Volume I, Concepts and Options for Determining Energy and Water Savings, effective 2001, Efficiency Valuation Organization (EVO)

2-5 pts

RESPONSIBLE PARTY:
MECHANICAL ENGINEER

EA Credit 3:
Measurement and Verification

ANSWER KEY

PURPOSE

To provide the opportunity to track the tenant's **energy** and **water** consumption over time.

REQUIREMENTS

Case 1: Tenants that lease **less** than **75%** of total building area (complete at least **one** of the following):

- Install **submetering** equipment (2 points).
- Tenant **pays** for energy costs (not included in base rent) (5 points).

Case 2: Tenants that lease 75% or more of total building area (5 points):

- Install **continuous** metering equipment.
- Create and implement a **measurement** & verification (M&V) plan.
- Develop a plan for **corrective** action if expected energy savings are not achieved.

REFERENCED STANDARD

International Performance **Measure** & **Verification** Protocol (IPMVP), Volume I, Concepts and Options for Determining Energy and Water Savings, effective 2001, Efficiency Valuation Organization (EVO)

| CONSTRUCTION | EA Credit 4: Green Power |

PURPOSE

To reduce pollution generated by power plants by purchasing energy from _____ -source, _____ energy systems.

REQUIREMENTS

For _____ years, purchase _____ percent of the building's electrical demand from a Green-e _____ provider or purchase Green-e accredited renewable energy _____ (RECs).

Based on expected consumption determined by energy _____ from EA Credit 1.3: HVAC or default rate of _____ kilowatt-hours per square foot per year.

EXEMPLARY PERFORMANCE

For _____ years, purchase _____ percent of the building's electrical demand (or default rate of _____ kilowatt-hours per square foot per year) from a Green-e certified provider or purchase Green-e accredited renewable energy certificates (RECs).

REFERENCED STANDARD

Center for Resource Solutions, _____ e Product Certification

5 pts

RESPONSIBLE PARTY:
OWNER

EA Credit 4:
Green Power

ANSWER KEY

PURPOSE

To reduce pollution generated by power plants by purchasing energy from **grid**-source, **renewable** energy systems.

REQUIREMENTS

For **two** years, purchase **50** percent of the building's electrical demand from a Green-e **certified** provider or purchase Green-e accredited renewable energy **certificates** (RECs).

Based on expected consumption determined by energy **model** from EA Credit 1.3: HVAC or default rate of **8** kilowatt-hours per square foot per year.

EXEMPLARY PERFORMANCE

For **two** years, purchase **100** percent of the building's electrical demand (or default rate of **16** kilowatt-hours per square foot per year) from a Green-e certified provider or purchase Green-e accredited renewable energy certificates (RECs).

REFERENCED STANDARD

Center for Resource Solutions, **Green**-e Product Certification

CHAPTER 7
MATERIALS AND RESOURCES

AS THE PREVIOUS CHAPTERS POINTED OUT, the built environment can be quite tolling on the natural environment. This book has so far presented means of minimizing impacts from the project site and reducing water and energy demands, while this chapter details strategies to minimize the environmental impacts of building materials as depicted in the Materials & Resources (MR) category in the Leadership in Energy and Environmental Design (LEED®) for Commercial Interiors ™(CI) rating system (Figure 7.1). This chapter details how to properly select materials and what to do with them after their useful life. These are two critical elements for the environment and the building industry, as buildings are a large consumer of natural resources and also contribute to the amount of solid waste generated, not only from an operational standpoint but also in terms of construction. More specifically, a sustainability guide by San Mateo County suggests, "Construction in the United States consumes 25 percent of all wood that is harvested, 40 percent of all raw stone, gravel and sand."[1] In terms of the quantity of waste generated, "according to the U.S. Environmental Protection Agency (EPA), an estimated 76 million tons of debris was generated in 1996 from commercial construction projects, including renovation and demolition activities."[2] As a result, green building project team members are advised to evaluate the environmental impact of their materials and product specifications and how to address waste during construction and operations.

It's time to pick a different color for flashcards created for MR topics.

Project teams may find themselves asking, "Where does steel come from? What kinds of materials are used to make green building products? How far did the raw material for the windows have to travel to the manufacturing plant? How far is the manufacturing plant from the project site? What happens to the leftover gypsum wallboard scraps? "Can we reduce the amount of waste leaving a project site? Is there a way to reduce the amount of packaging used to transport materials and products?" To help answer these types of questions, this chapter addresses three components for consideration as related to material and resource selection and disposal:

1. Salvaged materials and material reuse
2. Building material selection
3. Waste management

Each of these strategies is referred to in the prerequisites and credits of the category, as seen in Table 7.1. Remember from the previous chapters to

Create a flashcard to remember the three components to address within the MR category.

Figure 7.1 Specifying green materials, such as these 3Form® panels with recycled content, is a strategy to reduce the detrimental impacts of construction and the need for virgin materials.
Photo courtesy of Skylar Nielson, 3Form

note when the prerequisite or credit is eligible to be submitted for review: at the end of design or construction. Notice that only the prerequisite and Credit 1.1 can be submitted after construction documents (CDs) are completed, while all of the other credits are submitted for certification review after substantial completion. Do not worry about remembering the exact credit name and number, but it is important to know which are credits versus prerequisites.

Table 7.1 The Materials & Resources Category

D/C	Prereq/Credit	Title	Points
D	Prereq 1	Storage and Collection of Recyclables	R
D	Credit 1.1	Tenant Space, Long-term Commitment	1
C	Credit 1.2	Building Reuse, Maintain Interior Nonstructural Elements	1 to 2
C	Credit 2	Construction Waste Management	1 to 2
C	Credit 3.1	Materials Reuse	1 to 2
C	Credit 3.2	Materials Reuse, Furniture and Furnishings	1
C	Credit 4	Recycled Content	1 to 2
C	Credit 5	Regional Materials	1 to 2
C	Credit 6	Rapidly Renewable Materials	1
C	Credit 7	Certified Wood	1

SALVAGED MATERIALS AND MATERIAL REUSE

Material selection should begin with the discovery of what is already existing that can be reused. From a large-scale perspective, reusing existing building stock can reduce the burden on the infrastructure, such as utility and road expansion (Figure 7.2). Reusing building components reduces the demand for virgin materials and the impacts from manufacturing, and can add to the character of a project (Figure 7.3). It also helps to reduce waste, as these salvaged materials are reused as they are or **adaptively reused** (repurposed); in either case, they are diverted from landfills and incineration facilities and eliminate the need for packaging that is discarded. The *ID+C Reference Guide* addresses reuse in three ways to encourage pursuit by project teams.

Figure 7.2 Finding new uses for existing buildings helps to extend the life of the existing building stock and avoids demolition and waste. *Photo courtesy of SmithGroup, Inc.*

Figure 7.3 The ebony oak wood for the reception desk at the One Haworth Center in Holland, Michigan, was recovered from the Great Lakes and other waterways to avoid depleting old-growth forests.
Photo courtesy of Haworth Inc.

 TIP Materials are calculated by square footage for contribution. For interior doors and exterior walls, count one surface side, but for interior walls, include the square footage for both surfaces. Approach the calculations as if you were a painting contractor and bidding on the surface areas to be painted.

 TIP MR Credit 1.2 compliance is based on the square footage of the total preserved existing area in relation to the new square footage implemented.

 TIP Reused materials are calculated as a percentage of total material cost of a project to prove compliance with MR Credit 3.1. Therefore, if a project team implements $50,000 worth of salvaged, reused, and/or **refurbished** materials on a project with a total material cost of $1 million, the project would earn 1 point.

Salvaged Materials and Material Reuse in Relation to LEED Compliance

MR Credit 1.2: Building Reuse, Maintain Interior Nonstructural Elements. Project teams that preserve at least 40 percent (by area) of interior components, such as built-in casework, doors, ceilings, partitions, flooring and walls, are eligible to pursue this credit and earn one point. To earn another point, 60 percent must be preserved. To be eligible for inclusion for this credit's calculations, these items must be used as they were originally intended. Materials that do not meet the intentions of this credit can be used in the calculations for MR Credit 2: Construction Waste Management.

MR Credit 3.1: Materials Reuse. This credit differs from the previous, as it allows materials to be salvaged on-site and also from off-site. The qualifying attributes are determined by the intended use of the material to be reused. For example, if the material is found on-site, it must be repurposed from its original intention, such as a door to be reused as a table, and cannot be used in the calculations for any other MR credit. Remember, those materials serving their intended function and found on-site are to be used toward the previously described credits. If the material is salvaged from another location, it can serve the same purpose or have a new one to qualify with the intentions of this credit. Any off-site, salvaged materials implemented on a project can also be applied toward MR Credit 5: Regional Materials to be discussed later in the chapter. The minimum threshold to earn this credit is 5 percent (based on cost) to earn 1 point and 10 percent to earn 2 points. The rating system offers the opportunity to earn an Exemplary Performance credit should 15 percent of the total materials purchased is considered salvaged, reused, or refurbished. Mechanical, electrical, and plumbing fixtures and equipment must be excluded from the calculations. Products that fall into Construction Specification Institute (CSI) MasterFormat™ Divisions 03–10, plus

Foundations and Sections, Paving, Site Improvements, and Plantings (Divisions 31 and 32) should be included in the compliance calculations.

MR Credit 3.2: Materials Reuse, Furniture and Furnishings. This credit differs from the previous, as it addressing furniture to be reused or refurbished. To be eligible the furniture can be purchased from a used furniture supplier or can be part of the tenant's existing inventory. The minimum threshold to earn this credit is 30 percent of the furniture budget (based on cost) to earn one point. The rating system offers the opportunity to earn an Exemplary Performance credit should 60 percent of the total furniture budget be considered salvaged, used, or refurbished.

Create a flashcard to remember the three credit opportunities that address salvaged materials and material reuse in the LEED CI rating system.

QUIZ TIME!

Q7.1. In order to comply with MR Credit 3.1: Materials Reuse, how many points could a project seeking LEED for Commercial Interiors certification earn if the contractor purchased $65,000 of salvaged lumber from a nearby facility and $55,000 of a refurbished raised access floor system? The team is planning on reusing the existing doors, with a value of $41,000, although they are relocating them to new locations throughout the tenant space. The total material cost for the project has a value of $2 million. (Choose one)

  A. One
  B. Two
  C. This project does not meet the intentions of the credit.
  D. Three, including one for exemplary performance
  E. Four, including one for exemplary performance and Innovation in Design

Q7.2. Which of the following credits could a project team pursue if the contractor purchased refurbished demountable partitions from a nearby supplier? (Choose two)

  A. MR Credit 4: Recycled Content
  B. MR Credit 5: Regional Materials
  C. MR Credit 3.1: Material Reuse
  D. MR Credit 3.2: Material Reuse, Furniture and Furnishings
  E. MR Credit 1.2: Building Reuse, Maintain Interior Nonstructural Elements

Q7.3. Which of the following credits offer exemplary performance opportunities? (Choose three)

  A. SS Credit 2: Development Density and Community Connectivity
  B. MR Credit 3.1: Material Reuse
  C. EA Credit 3: Measurement and Verification
  D. MR Credit 1.2: Building Reuse, Maintain Interior Nonstructural Elements
  E. WE Credit 1: Water Use Reduction

Q7.4. Which of the following are excluded from compliance with MR Credit 1.2: Building Reuse, Maintain Interior Nonstructural Elements? (Choose two)

  A. Window assemblies
  B. Doors
  C. Interior partitions
  D. Ceiling
  E. Hazardous materials

Q7.5. Which of the following statements are true? (Choose two)

  A. Compliance with MR Credit 3.1: Materials Reuse is based on purchasing salvaged materials with a value of at least 5 percent of the total material cost for the project.
  B. Mechanical, electrical, and plumbing components cannot be salvaged and reused.
  C. Existing structural building elements that are maintained can be included in all of the MR credits.
  D. Hazardous materials can be included in the MR credit calculations, as long as they are remediated to the ASTM E-1903 Standard.

MATERIAL SELECTION

Implementing sustainable building materials affects a project's triple bottom line, just as with site selection and energy and water demands. As introduced in Chapter 2, project teams should perform **life-cycle assessments (LCAs)** of building materials, prior to specification, to evaluate the "cradle-to-grave" cycle of each material, especially as related to the environmental components of pollution and the demand of natural resources. The cradle-to-grave cycle includes the **embodied energy,** such as the extraction location of raw materials, the manufacturing process and location, the impact on construction workers and building occupants, the expected term of use during operations, the disposal options available, and the energy contained within the product itself. With the evaluation of these components, the results of an LCA will help to determine the material selections to include in the construction purchasing policy to help guide the contractor.

Once the materials are selected, they should be evaluated for compliance with LEED. The credit opportunities will be presented in the next section, but first it is important to decipher how each is calculated and what should be included in the calculations. In the previous section, compliance with MR Credit 1.2 is based on area, whereas Credit 3.1 calculations are based on a percentage of the total material cost of the project and Credit 3.2 is based on total furniture and furnishings budget. For existing materials to be reused, such as furniture or door hardware, a replacement cost would need to be determined.

How does a team determine the total material cost of a project? Project teams will need to calculate the actual cost of materials (not including labor or equipment) included in CSI MasterFormat Divisions 03–10, plus Foundations and

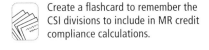
Create a flashcard to remember the CSI divisions to include in MR credit compliance calculations.

Sections, Paving, Site Improvements, and Plantings. Mechanical, electrical, and plumbing equipment and any specialty items, such as elevators, are not included in any of the calculations.

Once the total material cost is determined for the project, the products are required to be tracked, measured, and calculated to prove compliance. Most of the MR credits are calculated in a similar way; for example, recycled content, regional materials, salvaged materials, and rapidly renewable animal or fiber products are calculated as a percentage of the total material cost for a project. Forest Stewardship Council (FSC) wood products are calculated as a percentage of the total cost of new wood products purchased for a specific project. Remember, the previously presented building reuse credit is based on area, not on cost. After construction, materials are documented and tallied to show compliance to earn points. For example, should a project purchase 60 percent of their new wood products from a sustainably managed forest, they would earn one point, as the minimum threshold requires the team to purchase at least 50 percent FSC-certified wood of the total new wood purchased.

Project teams may be presented with a building component assembled with multiple materials. Each of the different materials could be manufactured with different environmental impacts, such as extraction and processing locations, and they can contain different amounts of postconsumer recycled content, preconsumer recycled content, or none at all. In this case, the team would need to evaluate the **assembly recycled content** of the component. It does not matter if the assembly consists of multiple materials (think concrete) or of multiple subcomponents (think furniture). To calculate compliance with MR Credit 4: Recycled Content, teams would need to divide "the weight of the recycled content by the overall weight of the assembly."[3] For example, if the window assemblies were assessed, the frame could have a different recycled content than the glass itself. The weight of the frame would need to be determined versus the weight of the glass, and then tallied with proportionate value of different types of recycled contents in order to determine the overall recycled content value of the assembled material. For assembly products with different recycled contents values, the recycled contents have to be broken out and multiplied by the cost. For example, if a toilet partition has 30 percent preconsumer and 10 percent postconsumer recycled content values, and is worth $10,000, the responsible party would perform the following calculations:

$$\text{Preconsumer value of 30 percent: } \$10{,}000 \times 30\% \times .5 = \$1{,}500$$
$$\text{Postconsumer value of 10 percent: } \$10{,}000 \times 10\% = \$1{,}000$$
$$\text{Total product value contributing to MR Credit 4} = \$2{,}500$$

The same concept can be applied to other MR Credits, such as Material Reuse, Regional Materials, Rapidly Renewable Materials, and Certified Wood. For example, if the project team purchased a conference room table with a salvaged base and a new glass top, the weight of the base would need to be determined as a percentage of the table as a whole. If the base weighed 75 percent of the total assembly, 75 percent of the assembly cost could be used toward MR Credit 3.2 and possibly MR Credit 5 if the base was salvaged locally. Therefore, it may be necessary to have manufacturers to provide their product's assembly information categorized by weight.

Create a flashcard to remember how compliance is calculated for each applicable MR credit (area or cost). Remember MR Credit 7 is based only in wood costs and not total material costs.

Material Selection in Relation to LEED Compliance

Although it may not be feasible to conduct a full LCA for every product, project teams can refer to the *ID+C Reference Guide* for material selection assistance. The LEED CI rating system suggests that project teams implement products with one or more of the following characteristics (see Table 7.2).

MR Credit 4: Recycled Content. It is important to remember that this credit addresses the purchasing of materials with recycled content (Figure 7.4), not the actual recycling of materials, as that process will be addressed later. For project teams to comply with this credit, they will need to ensure that the contractor purchases materials with preconsumer and/or postconsumer recycled content equivalent to the cost of at least 10 percent of the total materials (based on cost) to be awarded one point. Table 7.2 defines the difference between the two types of recycled content and also provides examples of each. Project teams are eligible to earn 2 points if the amount reaches 20 percent and can earn a third for exemplary performance for purchasing 30 percent.

MR Credit 5: Regional Materials. Purchasing products from vendors nearby reduces transportation impacts, such as pollution, and also preserves the local economy (Figure 7.5). But what is considered local? This credit requires a project team to track not only the location from which a product was sold or distributed but also the extraction points of the materials to make the product, where those materials were processed, and where the final product was manufactured and assembled. Each of these location points have to be determined for each of the materials contributing to earning this credit. In order to earn one point, a project team would need to demonstrate that at least 20 percent of the combined value of material cost (including Division 12 (furniture and furnishings) were manufactured locally. If 10 percent of materials were extracted, processed, *and* manufactured within 500 miles of the project site, the project

 Create a flashcard to remember ISO 14021-1999: Environmental Label and Declarations is the referenced standard that declares a material having postconsumer/preconsumer recycled content.

 TIP Steel is the only material with a default recycled content value, as sometimes there is no documentation available to prove compliance. Project teams are then able to apply a 25 percent postconsumer value for any steel products implemented on the project when calculating the steel product's contribution toward earning MR Credit 4.

 Remember, salvaged materials recovered from within 500 miles of the project site can contribute toward earning either of the MR Credit 3 opportunities and MR Credit 5 as well. The recovery location is the extraction point and the point of sale acts as the manufacturing location.

Figure 7.4 Permeable pavers made with recycled content not only help to recharge the groundwater but also help to reduce the need for virgin materials. *Photo courtesy of Vast Enterprises, LLC*

Figure 7.5 Purchasing materials that are extracted, processed, and manufactured within 500 miles helps to reduce the transportation impacts associated with building materials. These Icestone products have 26 percent (by weight) material extracted from York, Pennsylvania, and 100 percent of the product manufactured in Brooklyn, New York. *Photo courtesy of Icestone, LLC and Green Team*

team can pursue another point within this credit. Teams earn an Exemplary Performance credit for purchasing 20 percent of the total materials for the project (by cost) that were extracted, processed, and manufactured within 500 miles of the project site.

As mentioned previously, for assembled materials with only a portion of the components that comply, that qualifying portion is calculated by weight for the purposes of documentation.

 Create a flashcard to remember that MR Credit 5 offers two point opportunities with the differentiator being the extraction location from the project site.

Create a flashcard to remember that rapidly renewable fiber or animal materials must be grown or raised in 10 years or less.

Which CSI Divisions are to be included in the MR calculations? Make a flashcard to recall them if you do not remember!

MR Credit 6: Rapidly Renewable Materials. Rapidly renewable materials are "agricultural products, both fiber and animal, that take 10 years or less to grow or raise and can be harvested in a sustainable fashion"[4] and, therefore, preserve natural resource materials for future generations. Project teams that purchase rapidly renewable materials worth 5 percent of the total material cost of the project are eligible to pursue this credit. If they purchase 10 percent or more, they can earn an Exemplary Performance credit. Be sure to refer to Table 7.2 for examples of rapidly renewable materials, and note that both animal and fiber types are included.

MR Credit 7: Certified Wood. The **Forest Stewardship Council (FSC)** certification proves compliance for responsible forest management, preserving materials for future generations and habitats, as well as maintaining biodiversity (Figure 7.6). Teams must purchase at least 50 percent of the new, permanently installed wood for the project that is FSC certified and collect the vendor's or manufacturer's **chain-of-custody (COC)** certification documentation to prove compliance in order to earn this credit. Remember, this credit is calculated with only the new wood purchased for the project and *not* the total material cost, and it does not include temporarily installed wood, such as formwork. It can, however, include different types of FSC-certified products, such as FSC Pure, FSC Mixed Credit, and FSC Mixed (NN) percent. For the purposes of calculating compliance, FSC Pure and FSC Mixed Credit are valued at 100 percent of their cost, while FSC Mixed (NN) percent shall only have a value of its percentage of FSC product. For example, wood that is certified FSC Mixed (35) percent can contribute only 35 percent of its value toward compliance. FSC Recycled certified products do not meet the intentions of this credit, but they can be included in the MR Credit 4: Recycled Content calculations. Should project teams purchase more than 95 percent

Figure 7.6 Purchasing wood from sustainable and responsible forests helps to ensure resources for future generations.
Photo courtesy of Armstrong Ceiling and Wall Systems

of FSC certified wood, the project is eligible for an Exemplary Performance credit under the Innovations in Design (ID) category.

Table 7.2 Green Building Products

Characteristic		Description	Examples
Materials with recycled content		Products manufactured with material previously used	Masonry, concrete, carpet, acoustic ceiling tile, tile, rubber flooring, insulation, metal, and gypsum wallboard
	Preconsumer waste	Material left over from the manufacturing process and implemented into a new manufacturing process	Fly ash, sawdust, walnut shells, sunflower seed hulls, obsolete inventories, shavings, and trimmings
	Postconsumer waste	Manufactured products at the end of their useful life	Any products that were consumed (such as metals, plastics, paper, cardboard, glass), and demolition and construction debris
Local/regional materials		Products that are extracted, processed, and manufactured close to a project site	Materials obtained within 500 miles of the project site
Rapidly renewable materials		Animal or fiber materials that grow or can be raised in less than 10 years	Bamboo flooring and plywood, cotton batt insulation, linoleum flooring, sunflower seed board panels, wheatboard cabinetry, wool carpeting, cork flooring, bio-based paints, geotextile fabrics, soy-based insulation, and straw bales
FSC-certified wood materials		Sustainably managed forest resources	Contractors are required to show chain-of-custody (COC) documentation

 Create a flashcard to remember that FSC wood requires chain-of-custody (COC) documentation.

 TIP What types of green building materials are calculated the same for LEED documentation purposes? Which ones are calculated as a percentage of the total material volume or weight?

 TIP Fly ash can be a substitution for Portland cement for concrete products. It is the residual component that is left behind during the coal incineration or combustion process.

 Create a flashcard to remember the difference between preconsumer and postconsumer recycled contents.

 Do you remember what form the contractor would fill in to upload to LEED-Online to show compliance with the recycled content, local/regional materials, FSC wood, or rapidly renewable credits?

QUIZ TIME!

Q7.6. Which of the following materials could contribute to earning MR Credit 4: Recycled Content credit, as preconsumer recycled content? (Choose three)

　A. Metal stud manufacturing scrap sent back into the same manufacturing process

　B. Paper towels manufactured from cardboard used for packaging

　C. Medium-density fiberboard panels manufactured with sawdust generated by the manufacturing of structural insulated panels

　D. Concrete made with fly ash collected from coal-burning power plants

　E. Carpet padding manufactured with waste fiber collected from textile manufacturing plants

Q7.7. The percentage calculation for MR Credit 6: Rapidly Renewable Materials accounts for which of the following? (Choose two)

   A. Cost of rapidly renewable materials
   B. Volume of rapidly renewable materials
   C. Combined weight for all rapidly renewable materials
   D. Total materials cost for the project

Q7.8. Which of the following is not an example of a rapidly renewable material? (Choose one)

   A. Strawboard
   B. Oak wood flooring
   C. Cotton insulation
   D. Cork flooring
   E. Wheatboard

Q7.9. A tenant purchased FSC-certified doors and trim. She is interested in determining whether the project meets the requirements of MR Credit 6: Certified Wood for LEED compliance. The qualification of products and determination of their contribution to certified wood requires keeping track of which of the following? (Choose two)

   A. Cost of certified wood products as a percentage of the total material cost of the project
   B. Cost of certified wood products as a percentage of the total cost for wood products purchased for the project
   C. Weight of certified wood products as a percentage of all new wood products used on the project
   D. Chain-of-custody documentation for all FSC-certified wood products
   E. Chain-of-custody documentation for all new wood products purchased for the project

Q7.10. To help calculate compliance for MR credits, what is the default percentage value project teams can use to determine the total material cost for a project? (Choose one)

   A. 10 percent
   B. 20 percent
   C. 25 percent
   D. LEED for Commercial Interiors does not utilize a default value for calculating compliance.
   E. 35 percent

Q7.11. Which of the following statements are not true? (Choose three)
　A. Furniture must be included in the calculations for MR credits.
　B. Specialty equipment cannot be included in the calculations for MR credits.
　C Furniture cannot be included in the calculations for MR credits.
　D Mechanical, electrical, and plumbing equipment and fixtures cannot be included in the calculations for MR credits.
　E. Mechanical, electrical, and plumbing equipment and fixtures must be included in the calculations for MR credits.
　F. Specialty equipment must be included in the calculations for MR credits.

Q7.12. Which of the following statements are true? (Choose two)
　A. FSC is the acronym for Forest Stewardship Council.
　B. FSC-certified wood and rapidly renewable materials can also count toward MR Credit 5: Regional Materials if they are extracted, processed, and manufactured within 500 miles of the project site.
　C. FSC-certified wood and rapidly renewable materials cannot count toward MR Credit 5: Regional Materials.
　D. FSC-certified wood and rapidly renewable materials can also count toward MR Credit 5: Regional Materials if they are extracted, processed, and manufactured within 100 miles of the project site.
　E. FSC is the acronym for Federal Society of Certified wood.

WASTE MANAGEMENT

Construction processes and building operations should be addressed to minimize environmental impacts from disposal and waste. In the United States, building construction and demolition alone account for 40 percent of the total waste stream, while building operations account for 300 tons of waste per year for a building with 1,500 employees.[5]

When waste is collected and hauled from a facility, it is typically brought to a landfill or an incineration plant, both of which contribute to greenhouse gas emissions. Landfills produce and then leak methane, and incineration facility processes produce carbon dioxide. As another environmental detriment, think about the potential for landfills to contaminate groundwater sources. As a result, green building project teams and facility managers are encouraged to address waste diversion strategies for projects under construction and operational buildings to avoid landfills and incineration facilities (Figure 7.7).

 TIP Landfills require sunlight, moisture, and oxygen in order to decompose material—quite a challenging feat for a dark, enclosed environment, don't you think?

160 CHAPTER 7: MATERIALS AND RESOURCES

Create a flashcard to remember the EPA statistic for current recycling rates of 32 percent.

Create a flashcard to remember the 3R's of waste management: Reduce, Reuse, and Recycle. Remember them in that order as a hierarchal approach for policies. Source reduction is at the top because it "reduces environmental impacts throughout the material's life-cycle, from the supply chain and use to recycling and waste disposal."[7] Source reduction should be remembered as the effort to reduce the amount of material brought into a facility.

Create a flashcard to remember the minimum types of items to be recycled during operations to meet the requirements of the MR prerequisite: paper, corrugated cardboard, glass, plastics, and metals.

Figure 7.7 Separating waste on-site is one method to comply with the construction waste management strategies to avoid using landfills and incineration facilities.
Photo courtesy of Auld & White Constructors, LLC

The EPA estimates a reduction of 5 million metric tons of carbon dioxide if recycling efforts were to increase just 3 percent above the current 32 percent rate.[6] To help reach this goal, the LEED CI rating system offers point opportunities for implementing waste management policies during construction, to divert waste by reuse and recycling strategies.

Construction waste management plans should address whether waste will be separated on-site into individually labeled waste containers or collected in a **commingled** fashion in one container and sorted off-site (Figure 7.7). As with many of the components addressed within the LEED rating systems, there are trade-offs when deciding between the two options. Commingled collection reduces the amount of space needed on-site, while on-site collection may require additional labor to manage the sorting effort. In either case, land-clearing debris and soil should not be included in the calculations, but metals, concrete, and asphalt should all be collected for recycling and accounted for (Figure 7.8). Recycling options for paper, cardboard, plastics, and wood vary by region.

Waste Management in Relation to LEED Compliance

MR Prerequisite 1: Storage and Collection of Recyclables. Project teams are required to encourage recycling during operations by providing locations for collecting and storing materials to be recycled (Figure 7.9). At a minimum, the teams need to provide locations to collect paper, corrugated cardboard, glass, plastics, and metals. Calculations are not required to determine the size of the collection and storage locations, but the reference guide provides some guidance from the city of Seattle for the project teams to follow.

MR Credit 1.1: Tenant Space, Long-Term Commitment. Every time a tenant space is leased, typically the new tenant will modify the space to meet their needs. With this turnover, demolition and construction waste is generated. As a first effort to avoid future contributions to a landfill, this credit encourages

Figure 7.8 Dedicated waste container for masonry to be collected for recycling.

longer lease commitments to avoid less than a ten-year turnover. Project teams will need to provide a copy of the lease to prove compliance. Project teams may also comply should they own the space in which they reside.

MR Credit 2: Construction Waste Management. This credit is addressed during construction, whereas the MR prerequisite prepares the building for the process to be conducted during operations. For this credit, project teams are required to develop a construction waste management plan to be implemented by the contractor during construction to divert nonhazardous demolition and

Figure 7.9 Recycling during operations is required of all projects seeking LEED certification.
Photo courtesy of Jason Hagopian, AIA, LEED AP

TIP Waste is calculated in volume or weight (tons)

Create a flashcard to remember the three strategies to reduce waste.

construction debris from landfills and incineration facilities. The plan must describe the approach specific to each type of material and must instruct the contractor to collect the waste on-site either in a commingled or separated fashion. The plan will also indicate the need for the contractor to collect the receipts from the waste hauler to document proof of diversion. A team will earn one point for diverting 50 percent and another for diverting 75 percent of demolition and construction waste. They can pursue an exemplary performance opportunity for recycling 95 percent. The percentage of waste diverted can be calculated based on weight or volume, as long as it is done consistently throughout the construction process.

QUIZ TIME!

Q7.13. Which of the following greenhouse gases is a by-product of landfills? (Choose one)

   A. Carbon monoxide
   B. Methane
   C. Sulfur dioxide
   D. Nitrous oxide

Q7.14. Which of the following statements are not true about green building materials? (Choose two)

   A. Rapidly renewable materials are harvested within 10 years.
   B. Cradle-to-grave materials can be recycled.
   C. Products with postconsumer recycled content can contribute to earning MR Credit 4: Recycled Content, while products with preconsumer recycled content cannot.
   D. Cotton insulation can be considered a type of rapidly renewable material.

Q7.15. Which of the following would the contractor upload to LEED-Online? (Choose two)

   A. Stormwater management plan
   B. Total amount of waste diverted from a landfill
   C. The total material cost of the project and the percentage containing recycled content
   D. Energy modeling calculations to show expected energy savings

Q7.16. When addressing materials and resources, a project team should incorporate which of the following to comply with the credits and prerequisites of the MR category within the LEED CI rating system? (Choose three)

   A. The purchasing policy of the manufacturer
   B. Location of manufacturing plant of steel
   C. The type of car the CEO drives
   D. Postconsumer recycled content of a chair
   E. Extraction location of silica

Q7.17. How would a contractor show proof of compliance with MR Credit 7: Certified Wood? (Choose one)

  A. Provide the architect with a spreadsheet summarizing all of the wood was purchased with 500 miles of the project site

  B. Upload receipts of FSC wood purchased to LEED-Online

  C. Upload chain-of-custody documentation to LEED-Online

  D. Complete a LEED credit form and upload it to LEED-Online including the chain-of-custody numbers and invoice amounts

  E. Mail all documentation to USGBC and GBCI

Q7.18. How is waste hauled from a construction site calculated for the purposes of LEED? (Choose one)

  A. As a percentage of total material cost of a project

  B. As a percentage of the total material volume or weight

  C. As a percentage of total cost of a project

  D. In tons

Q7.19. Which of the following are not required for collection to comply with MR Prerequisite 1: Storage and Collection of Recyclables? (Choose two)

  A. Glass
  B. Plastics
  C. Wood
  D. Rubber
  E. Corrugated cardboard
  F. Metal

Q7.20. A contractor was provided with receipts from his waste hauler indicating that 20 tons of wood, plastic, and metal waste were sent to manufacturers for reuse. The hauler also indicated that another 20 tons were landfilled, while 15 tons were sent to an incineration plant and 8 tons of soil and plant debris were sent to a site in need of mulching material. The hauler's receipts also show that 45 tons of concrete were sold to a concrete masonry unit manufacturing facility over 500 miles away from the project site. What percentage of waste was diverted in order to comply with MR Credit 2: Construction Waste Management? (Choose one)

  A. 55 percent
  B. 65 percent
  C. 70 percent
  D. 62.2 percent

DESIGN

**MR Prerequisite 1:
Storage and Collection of Recyclables**

PURPOSE

Reduce _____ generated by building occupants that is _____ to and _____ of in _____.

REQUIREMENTS

Allocate space for the _____ and _____ of recyclable materials within the entire building.

At a minimum, collect _____, corrugated cardboard, _____, _____, and _____.

DOCUMENTATION

Provide a _____ indicating storage and collection areas.

Provide a copy of the building's _____ program.

Required

RESPONSIBLE PARTY:
OWNER

	MR Prerequisite 1: Storage and Collection of Recyclables
ANSWER KEY	
PURPOSE	Reduce **waste** generated by building occupants that is **hauled** to and **disposed** of in **landfills**.
REQUIREMENTS	Allocate space for the collection and storage of recyclable materials within the entire building. At a minimum, collect **paper**, corrugated cardboard, **glass**, **plastics**, and **metals**.
DOCUMENTATION	Provide a **plan** indicating storage and collection areas. Provide a copy of the building's **recycling** program.

DESIGN

MR Credit 1.1:
Tenant Space – Long-Term Commitment

PURPOSE

Encourage tenants to _____ resources, reduce waste, and reduce the environmental impacts of leased spaces as related to _____, manufacturing, and _____.

REQUIREMENTS

Commit to lease a space for at least _____ years.

1 pt

RESPONSIBLE PARTY:
TENANT

MR Credit 1.1:
Tenant Space - Long-Term Commitment

ANSWER KEY

PURPOSE

Encourage tenants to **conserve** resources, reduce waste, and reduce the environmental impacts of leased spaces as related to **materials,** manufacturing, and **transport**.

REQUIREMENTS

Commit to lease a space for at least **10** years.

	MR Credit 1.2:
CONSTRUCTION	**Building Reuse, Maintain Interior Nonstructural Elements**

PURPOSE

To reduce the need for and the _____ and _____ impacts of new building materials by _____ the life cycle of the _____ building stock, _____ resources, retain _____ resources, reduce _____ and the environmental impacts of new buildings.

REQUIREMENTS

Preserve at least _____ percent of the existing interior nonstructural elements by _____ (1 point).

Preserve at least _____ percent of the existing interior _____ elements by area (2 points)

Interior nonstructural elements include walls, doors, _____ coverings, and _____ systems.

EXEMPLARY PERFORMANCE

Preserve at least _____ percent of the existing interior nonstructural elements by area.

1 pt

RESPONSIBLE PARTY:

ARCHITECT

MR Credit 1.2:
Building Reuse,
Maintain Interior Nonstructural Elements

ANSWER KEY

PURPOSE

To reduce the **need** for and the **manufacturing** and transportation impacts of new building materials by **extending** the life cycle of the **existing** building stock, conserve resources, retain **cultural** resources, reduce **waste** and the environmental impacts of new buildings.

REQUIREMENTS

Preserve at least **40** percent of the existing interior nonstructural elements by **area** (1 point).

Preserve at least **60** percent of the existing interior **nonstructural** elements by area (2 points)

Interior nonstructural elements include walls, doors, **floor** coverings, and **ceiling** systems.

EXEMPLARY PERFORMANCE

Preserve at least **80** percent of the existing interior nonstructural elements by area.

CONSTRUCTION

MR Credit 2:
Construction Waste Management

PURPOSE

_____ construction and _____ waste from _____ and _____ facilities and redirect the _____ material back into the manufacturing process or appropriate sites.

REQUIREMENTS

Create and implement a _____ _____ management plan.

Divert a percentage of construction and demolition debris by _____, _____, or _____.

POINT DISTRIBUTION

- 50% → 1 point
- 75% → 2 points
- 95% → 1 EP Point

DOCUMENTATION

Construction Waste Management Plan

Completed Submittal Form on LEED-Online with the following information:

- _____ description
- _____ of each type of waste (in terms of _____ or _____)
- _____ of disposal for each type of waste
- Percentage of total waste _____

EXEMPLARY PERFORMANCE

Divert more than _____ percent of construction waste from landfills and incineration facilities.

1–2 pts

RESPONSIBLE PARTY:
CONTRACTOR

MR Credit 2: Construction Waste Management

ANSWER KEY

PURPOSE

Divert construction and **demolition** waste from **landfills** and **incineration** facilities and redirect the **reusable** material back into the manufacturing process or appropriate sites.

REQUIREMENTS

Create and implement a **construction waste** management plan.

Divert a percentage of construction and demolition debris by **recycling**, **donation**, or **salvaging**.

POINT DISTRIBUTION

- 50% • 1 point
- 75% • 2 points
- 95% • 1 EP Point

DOCUMENTATION

Construction Waste Management Plan

Completed Submittal Form on LEED-Online with the following information:

- **Waste** description
- **Quantity** of each type of waste (in terms of **weight** or volume)
- **Destination** of disposal for each type of waste
- Percentage of total waste **diverted**

EXEMPLARY PERFORMANCE

Divert more than **95** percent of construction waste from landfills and incineration facilities.

172 CHAPTER 7: MATERIALS AND RESOURCES

CONSTRUCTION

MR Credit 3.1: Materials Reuse

PURPOSE

Reduce the demand for _____ building materials and waste by _____ existing products, to reduce the impacts affiliated with the _____ and _____ of new materials.

REQUIREMENTS

Implement a percentage (based on cost) of _____, _____, or reused building materials and products.

CALCULATIONS

Materials from CSI _____ Divisions 03 to 10, _____, and 32 are to be included in calculations.

Furniture, mechanical, _____, and plumbing elements and specialty items are not eligible to contribute to earning this credit.

Items salvaged from on-site and off-site are eligible, but on-site items must serve a _____ purpose than originally intended.

POINT DISTRIBUTION

5% → 1 point
10% → 2 points
15% → 1 EP Point

EXEMPLARY PERFORMANCE

Implement at least _____ percent of total material cost, salvaged, refurbished, or reused materials.

1–2 pts

RESPONSIBLE PARTY: **ARCHITECT**

MR Credit 3.1: Materials Reuse

ANSWER KEY

PURPOSE

Reduce the demand for **virgin** building materials and waste by **reusing** existing products, to reduce the impacts affiliated with the **extraction** and **processing** of new materials.

REQUIREMENTS

Implement a percentage (based on cost) of **salvaged**, **refurbished**, or reused building materials and products.

CALCULATIONS

Materials from CSI **MasterFormat**™ Divisions 03 to 10, **31**, and 32 are to be included in calculations.

Furniture, mechanical, **electrical**, and plumbing elements and specialty items are not eligible to contribute to earning this credit.

Items salvaged from on-site and off-site are eligible but on-site items must serve a **new** purpose than originally intended.

POINT DISTRIBUTION

- 5% • 1 point
- 10% • 2 points
- 15% • 1 EP Point

EXEMPLARY PERFORMANCE

Implement at least **15** percent of total material cost, salvaged, refurbished, or reused materials.

CONSTRUCTION

**MR Credit 3.2:
Materials Reuse,
Furniture and Furnishings**

PURPOSE

Reduce the demand for _____ materials and waste by _____ existing products, to reduce the impacts affiliated with the _____ and _____ of new materials.

REQUIREMENTS

Implement _____ percent (based on _____) of salvaged, _____, or reused furniture and furnishings based on the _____ furniture and furnishings budget.

EXEMPLARY PERFORMANCE

Implement _____ percent (based on _____) of _____ refurbished or _____ furniture and furnishings based on the _____ furniture and furnishings budget.

1 pt

RESPONSIBLE PARTY:

INTERIOR DESIGNER

MR Credit 3.2: Materials Reuse, Furniture and Furnishings

ANSWER KEY

PURPOSE

Reduce the demand for **virgin** materials and waste by **reusing** existing products, to reduce the impacts affiliated with the **extraction** and **processing** of new materials.

REQUIREMENTS

Implement **30** percent (based on **cost)** of salvaged, **refurbished,** or reused furniture and furnishings based on the **total** furniture and furnishings budget.

EXEMPLARY PERFORMANCE

Implement **60** percent (based on **cost)** of **salvaged,** refurbished, or **reused** furniture and furnishings based on the **total** furniture and furnishings budget.

CONSTRUCTION

MR Credit 4:
Recycled Content

PURPOSE
To increase the demand for products with _____ content to reduce the impacts from extracting and processing _____ materials.

REQUIREMENTS
Implement a percentage (based on cost) of materials and products with recycled content such that the sum of the _____ amount plus _____ of the _____ amount of content reaches the minimum percentage threshold.

CALCULATIONS
Materials from CSI MasterFormat™ Divisions 03 to _____, 12, 31, and 32 are to be included in calculations.

Mechanical, electrical, and _____ elements and specialty items are not eligible to contribute to earning this credit.

_____ must be includes in the calculations.

Recycled content of assembled materials is based on _____.

POINT DISTRIBUTION

- 10% → • 1 point
- 20% → • 2 points
- 30% → • 1 EP Point

EXEMPLARY PERFORMANCE
Implement at least _____ percent of total material cost, materials with recycled content.

REFERENCED STANDARD
International Standard _____ 14021-1999, Environmental Labels and Declarations, Self- _____ Environmental Claims (Type II Environmental Labeling)

1–2 pts

RESPONSIBLE PARTY:
CONTRACTOR

MR Credit 4: Recycled Content

ANSWER KEY

PURPOSE

To increase the demand for products with **recycled** content to reduce the impacts from extracting and processing **virgin** materials.

REQUIREMENTS

Implement a percentage (based on cost) of materials and products with recycled content such that the sum of the **postconsumer** amount plus **half** of the **preconsumer** amount of content reaches the minimum percentage threshold.

CALCULATIONS

Materials from CSI MasterFormat™ Divisions 03 to **10**, 12, 31, and 32 are to be included in calculations.

Mechanical, electrical, and **plumbing** elements and specialty items are not eligible to contribute to earning this credit.

Furniture must be included in the calculations.

Recycled content of assembled materials is based on **weight**.

POINT DISTRIBUTION

- 10% • 1 point
- 20% • 2 points
- 30% • 1 EP Point

EXEMPLARY PERFORMANCE

Implement at least **30** percent of total material cost, materials with recycled content.

REFERENCED STANDARD

International Standard **ISO** 14021-1999, Environmental Labels and Declarations, Self-**Declared** Environmental Claims (Type II Environmental Labeling)

CONSTRUCTION

MR Credit 5: Regional Materials

PURPOSE

To increase the demand for products that are _____ extracted and manufactured to reduce the impacts from _____ and to support _____ resources.

REQUIREMENTS

Option 1: (one point)

Implement at least _____ percent (based on cost) of materials and products that are _____ within _____ miles of the project site.

Option 2: (two points)

Implement at least _____ percent (based on cost) of materials and products that are _____, _____, and manufactured within _____ miles of the project site.

CALCULATIONS

Materials from CSI MasterFormat™ Divisions 03 to 10, _____, 31, and _____ are to be included in calculations.

Mechanical, electrical, and _____ elements and _____ items are not eligible to contribute to earning this credit.

_____ must be included in the calculations.

If only a portion of an assembled material is _____, _____, and _____ within 500 miles of the project site, the contributing value is based on _____.

EXEMPLARY PERFORMANCE

Implement at least _____ percent of total material cost, materials that are extracted, processed, and manufactured within _____ miles of the project site.

1-2 pts

RESPONSIBLE PARTY:
CONTRACTOR

MR Credit 5: Regional Materials

ANSWER KEY

PURPOSE

To increase the demand for products that are **locally** extracted and manufactured to reduce the impacts from **transportation** and to support **indigenous** resources.

REQUIREMENTS

Option 1: (one point)

Implement at least **20** percent (based on cost) of materials and products that are **manufactured** within **500** miles of the project site.

Option 2: (two points)

Implement at least **10** percent (based on cost) of materials and products that are **extracted**, **processed**, and manufactured within **500** miles of the project site.

CALCULATIONS

Materials from CSI MasterFormat™ Divisions 03 to 10, **12**, 31, and **32** are to be included in calculations.

Mechanical, electrical, and **plumbing** elements and **specialty** items are not eligible to contribute to earning this credit.

Furniture must be included in the calculations.

If only a portion of an assembled material is **extracted**, **processed**, and **manufactured** within 500 miles of the project site, the contributing value is based on **weight**.

EXEMPLARY PERFORMANCE

Implement at least **20** percent of total material cost, materials that are extracted, processed, and manufactured within **500** miles of the project site.

CONSTRUCTION

**MR Credit 6:
Rapidly Renewable Materials**

PURPOSE

To increase the demand for _____ _____ materials to reduce the use and depletion of finite raw materials and materials with a _____ growth cycle.

REQUIREMENTS

Implement at least _____ percent (based on cost) of materials and products that are able to be regrown or harvested in less than _____ years.

CALCULATIONS

Materials from CSI MasterFormat™ Divisions _____ to 10, 12, _____ , and 32 are to be included in calculations.

Mechanical, electrical, and plumbing elements and _____ items are not eligible to contribute to earning this credit.

Furniture must be included in the _____ .

If only a portion of an assembled material is rapidly renewable, the contributing value is based on _____ .

EXEMPLARY PERFORMANCE

Implement rapidly renewable materials that have a cost of at least _____ percent of the total material cost for the project.

1 pt

**RESPONSIBLE PARTY:
CONTRACTOR**

	MR Credit 6: **Rapidly Renewable Materials**
ANSWER KEY	
PURPOSE	To increase the demand for **rapidly renewable** materials to reduce the use and depletion of finite raw materials and materials with a **long** growth cycle.
REQUIREMENTS	Implement at least **5** percent (based on cost) of materials and products that are able to be regrown or harvested in less than **10** years.
CALCULATIONS	Materials from CSI MasterFormat™ Divisions 03 to **10**, 12, 31, and 32 are to be included in calculations. Mechanical, electrical, and plumbing elements and **specialty** items are not eligible to contribute to earning this credit. Furniture must be included in the **calculations**. If only a portion of an assembled material is rapidly renewable, the contributing value is based on **weight**.
EXEMPLARY PERFORMANCE	Implement rapidly renewable materials that have a cost of at least **10** percent of the total material cost for the project.

CONSTRUCTION

MR Credit 7:
Certified Wood

PURPOSE

To increase the demand for and encourage environmentally responsible _____ management.

REQUIREMENTS

Implement at least _____ percent (based on cost) of new wood products that are certified by the Forest _____ Council (FSC).

CALCULATIONS

Only _____ installed _____ wood materials (CSI Divisions 3–10 and 12) are to be included in calculations.

_____ must be included in the calculations.

Contractor will use the total _____ cost and not total _____ cost value when completing calculations.

Contractor will need to identify if wood products are FSC _____, FSC _____ Credit, or FSC Mixed (NN%).

DOCUMENTATION

_____ - of - _____ documentation

Invoices for _____ wood products purchased including percentage of certified wood.

EXEMPLARY PERFORMANCE

Implement at least _____ percent (based on cost) of new wood products that are certified by the Forest _____ Council (FSC).

REFERENCED STANDARD

_____ Stewardship _____'s Principles and Criteria

1 pt

RESPONSIBLE PARTY:
CONTRACTOR

	MR Credit 7: **Certified Wood**
ANSWER KEY	

PURPOSE

To increase the demand for and encourage environmentally responsible **forest** management.

REQUIREMENTS

Implement at least **50** percent (based on cost) of new wood products that are certified by the Forest **Stewardship** Council (FSC).

CALCULATIONS

Only **permanently** installed **new** wood materials (CSI Divisions 3-10 and 12) are to be included in calculations.

Furniture must be included in the calculations.

Contractor will use the total **wood** cost and not total **material** cost value when completing calculations.

Contractor will need to identify if wood products are FSC **Pure**, FSC **Mixed** Credit, or FSC Mixed (NN%).

DOCUMENTATION

Chain-of-**Custody** documentation

Invoices for **new** wood products purchased including percentage of certified wood.

EXEMPLARY PERFORMANCE

Implement at least **95** percent (based on cost) of new wood products that are certified by the Forest **Stewardship** Council (FSC).

REFERENCED STANDARD

Forest Stewardship **Council's** Principles and Criteria

CHAPTER 8
INDOOR ENVIRONMENTAL QUALITY

THIS CHAPTER FOCUSES ON THE ELEMENTS INVOLVED in improving the indoor environment as detailed in the Indoor Environmental Quality (EQ) category of the Leadership in Energy and Environmental Design (LEED®) rating systems. Remember that Chapter 2 introduced the importance of indoor environments, since Americans typically spend about 90 percent of their time indoors, according to the Environmental Protection Agency (EPA). The EPA also reports conventionally designed, constructed, and maintained indoor environments have significantly higher levels of pollutants than the outdoors.[1] However, studies have shown that green buildings with an improved interior environmental quality "have the potential to enhance the lives of building occupants, increase their resale value of the building, and reduce the liability for building owners."[2] Because employee salaries and benefits are typically the biggest cost for a business, larger than operating costs for facilities, such as utilities, the satisfaction and health of the occupants should be a high priority. Retaining employees in order to avoid the additional costs of training new hires can help to add efficiencies to the economic bottom line for businesses. Reducing absenteeism because of health impacts increases productivity and reduces liability of inadequate indoor environmental quality. Businesses, such as Haworth, will enjoy a return on their investment for increasing their employee satisfaction by providing a comfortable work environment (Figure 8.1). The LEED for Commercial Interiors™ (LEED CI) rating system offers the following strategies to address to improve indoor environments:

It's time to pick a different color for flashcards created for EQ topics.

1. Indoor air quality
2. Thermal comfort
3. Lighting (natural and artificial)

Create a flashcard to remember the three components of the EQ category.

Each of these strategies is referred to in the prerequisites and credits of the category, as seen in Table 8.1. Remember from the previous chapters to note when the prerequisite or credit is eligible to be submitted for review, at the end of design or construction. Notice only EQ Credits 3 and 4 must be submitted for review after substantial completion, while the prerequisites and all of the other credits can be submitted after construction documents are completed. Also notice most credits are only worth one point, with the exception of EQ Credit 8.1: Daylight and Views, Daylight. Do not worry about remembering the exact credit name and number, but it is important to know which are credits versus prerequisites.

CHAPTER 8: INDOOR ENVIRONMENTAL QUALITY

Figure 8.1 Addressing such factors as daylighting, views, and low-emitting materials helps to bring value to the indoor environmental quality. *Photo courtesy of Haworth Inc.*

Table 8.1 The Indoor Environmental Quality Category

D/C	Prereq/Credit	Title	Points
D	Prereq 1	Minimum Indoor Air Quality Performance	R
D	Prereq 2	Environmental Tobacco Smoke (ETS) Control	R
D	Credit 1	Outdoor Air Delivery Monitoring	1
D	Credit 2	Increased Ventilation	1
C	Credit 3.1	Construction IAQ Management Plan—During Construction	1
C	Credit 3.2	Construction IAQ Management Plan—Before Occupancy	1
C	Credit 4.1	Low-Emitting Materials—Adhesives and Sealants	1
C	Credit 4.2	Low-Emitting Materials—Paints and Coatings	1
C	Credit 4.3	Low-Emitting Materials—Flooring Systems	1
C	Credit 4.4	Low-Emitting Materials—Composite Wood and Agrifiber Products	1
C	Credit 4.5	Low-Emitting Materials—Furniture and Furnishings	1
D	Credit 5	Indoor Chemical and Pollutant Source Control	1
D	Credit 6.1	Controllability of Systems, Lighting	1
D	Credit 6.2	Controllability of Systems, Thermal Comfort	1
D	Credit 7.1	Thermal Comfort, Design	1
D	Credit 7.2	Thermal Comfort, Verification	1
D	Credit 8.1	Daylight and Views, Daylight	1 to 2
D	Credit 8.2	Daylight and Views, Views for Seated Spaces	1

INDOOR AIR QUALITY

The *LEED Reference Guide for Interior Design and Construction* (ID+C) uses the American Society of Heating, Refrigerating, and Air-Conditioning Engineers' (ASHRAE's) definition for **indoor air quality** (IAQ) as it states it "is the nature of air inside the space that affects the health and well-being of building occupants."[3] Studies have shown that poor indoor air quality can lead to respiratory disease, allergies and asthma, and **sick building syndrome** and can, therefore, affect the performance and productivity of employees. All of the LEED rating systems address components from a triple bottom line perspective, to improve air quality during construction and operations to avoid adverse effects on human health and to improve the quality of life. In order to achieve this, the rating systems require project teams to address IAQ in terms of ventilation rates, construction practices, and minimizing exposure by prevention and segregation.

Ventilation Strategies in Relation to LEED Compliance

Project teams are encouraged to provide adequate ventilation for occupants without compromising energy use efficiencies, not to be a burden on the environment by contributing to the need for fossil fuels. Mechanical systems should work to thermally balance outdoor air with every air change; therefore, the key is to find the right balance. Too many air changes are wasteful and would affect economic and environmental bottom lines. However, too little ventilation can result in reduced quality of the indoor air, which would affect the health and satisfaction of occupants, thus also affecting the triple bottom line components. Therefore, project teams designing green buildings use the industry standard **ASHRAE Standard 62.1, Ventilation for Acceptable for Indoor Air Quality**, to adequately and appropriately size mechanical systems that will deliver the proper amounts of outside air while balancing energy demands. ASHRAE 62 describes proper **ventilation rates,** or the "amount of air circulated through a space, measured in air changes per hour."[4]

 Increasing ventilation may improve the overall IAQ, but it may increase the energy demand of heating, ventilation, and air conditioning (HVAC) systems at the same time. Therefore, it is important to recognize this category focuses on the occupants' well-being and not energy efficiency.

EQ Prerequisite 1: Minimum Indoor Air Quality Performance. Just as with all the previous prerequisites described, this one sets a minimum level of performance required of all projects seeking LEED certification. The mechanical engineer should refer to ASHRAE 62.1-2007 or the local code (whichever is more stringent by requiring more outside air) to determine the outdoor air **ventilation rate** required to comply for a mechanically ventilated building. If the building has natural ventilation strategies, the mechanical engineer would consult paragraph 5.1 of ASHRAE 62.1. The standard addresses minimum requirements for ventilation rates in various building ventilation systems and types of occupied zones based on square footage, number of occupants and the associated activity, and the ventilation system itself. If the team is implementing a combination of natural and mechanical ventilation, or **mixed-mode ventilation,** these strategies are required to follow Chapter 6 of the referenced standard.

 Any time you see ASHRAE 62, think IAQ! Say it out loud IAQ 62, 62 IAQ!

Regardless of the ventilation strategy, if the project team is challenged to meet the minimum threshold of the referenced standard, the team will need to provide documentation of the reasoning, explaining why it is not possible to supply the minimum amount of outside air. An engineering assessment would also be required documenting the maximum ventilation, ensuring at least an absolute minimum of 10 cubic feet per minute per person is supplied.

188 CHAPTER 8: INDOOR ENVIRONMENTAL QUALITY

 TIP Be sure to remember that CO_2 sensors must be located between 3 and 6 feet above the floor (the air we breathe). CO_2 concentrations greater than 530 parts per million (ppm) as compared to exterior environments are typically harmful to occupants.

 Create a flashcard to remember **exfiltration** is the "air leakage through cracks and interstices and through the ceilings, floors, and walls."[5]

 Create a flashcard to remember the three ventilation strategies for improving indoor air quality.

EQ Credit 1: Outdoor Air Delivery Monitoring. The mechanical engineer will need to develop a strategy to include carbon dioxide (CO_2) sensors to ensure that minimum requirements are maintained to help promote occupant comfort and well-being. The strategy must include provisions for an alarm (visual or audible) to be generated if the CO_2 levels vary by 10 percent or more in **densely occupied spaces** (25 people per 1,000 square feet). Compliance strategies could include using the building automation system (BAS) to monitor CO_2 levels. For mixed-mode and naturally ventilated buildings, the BAS could work with the windows to automatically open should the CO_2 levels reach unhealthy limits. For projects without a connection to a BAS, an audible or visual alarm could indicate that the windows should be opened to increase ventilation (Figure 8.2). Mechanically ventilated buildings could incorporate demand-controlled ventilation where airflow is increased if maximum setpoints are exceeded.

EQ Credit 2: Increased Ventilation. Building off of Prerequisite 1, project teams with mechanical or mixed-mode ventilation systems must increase the outdoor air supply to exceed the referenced standard or local code by at least 30 percent in order to meet the intentions of this credit. For naturally ventilated project types, the teams are directed to comply with the Carbon Trust's *Good Practice Guide 237*. They will also need to complete the flow diagram process of Figure 1.18 as detailed in the Chartered Institution of Building Services Engineers (CIBSE) Applications Manual 10:2005, *Natural Ventilation in Non-Domestic Buildings*. The mechanical engineer has two compliance paths from which to choose. Option 1 requires diagrams and calculations to prove the strategy's effectiveness and compliance with the referenced standards, while option 2 requires an analytic model proving compliance with ASHRAE 62.1-2007, Chapter 6, for at least 90 percent of the occupied spaces. For the purposes of the exam, be sure to know how to comply, but not necessarily the details required to prove compliance, such as the details of the diagrams and calculations of CIBSE and the *Good Practice Guide 237*.

Figure 8.2 Providing operable windows offers building occupants access to means of adequate ventilation. *Photo courtesy of Gary Neuenswander, Utah Agricultural Experiment Station*

QUIZ TIME!

Q8.1. Which of the following EQ credits are to be submitted after substantial completion and are not eligible for a design-side review? (Choose two)

    A. Credit 4: Low-Emitting Materials

    B. Credit 2: Increased Ventilation

    C. Credit 3: Construction IAQ Management Plan

    D. Credit 1: Outdoor Air Delivery Monitoring

    E. Credit 5: Indoor Chemical & Pollutant Source Control

Q8.2. Which of the following statements is not true? (Choose one)

    A. All of the EQ credits are worth 1 point.

    B. There are two prerequisites within the EQ category.

    C. Although there are prerequisites within the EQ category, project teams only have to comply with at least one in order to pursue certification.

    D. EQ prerequisites and credits intend to improve occupant well-being and comfort and, therefore, productivity as well.

    E. EQ credits may compensate energy efficiency of the project, but improve the indoor air quality.

Q8.3. What percentage above the referenced standard must the mechanical engineer increase the ventilation rate in order to comply with EQ Credit 2: Increased Ventilation? (Choose one)

    A. 5 percent

    B. 10 percent

    C. 20 percent

    D. 30 percent

    E. 40 percent

Q8.4. What is considered a densely occupied area? (Choose one)

    A. At least 10 people per 100 square feet

    B. At least 25 people per 1,000 square feet

    C. At least 25 people per 10,000 square feet

    D. At least 25 people per 500 square feet

    E. At least 15 people per 1,000 square feet

Q8.5. In order to comply with EQ Credit 1: Outdoor Air Delivery Monitoring, what should the alarm set points be set to? (Choose one)

    A. +/− 1 percent

    B. +/− 3 percent

    C. +/− 5 percent

    D. +/− 10 percent

    E. +/− 15 percent

IAQ Practices during Construction in Relation to LEED Compliance

During construction, project teams should follow the practices recommended by the **Sheet Metal and Air Conditioning Contractors' National Association (SMACNA)** guidelines (Figure 8.3). Contractors need to be mindful of reducing or eliminating contaminants from entering the indoor environment, including mechanical systems, to deliver an environment with better air quality. Contaminants include **volatile organic compounds (VOCs)**, carbon dioxide, particulates, and tobacco smoke. SMACNA guidelines recommend that VOCs from furniture, paints, adhesives, and carpets should be kept below defined maximum levels to avoid polluting the indoor environment (Figure 8.4). For the purposes of LEED, the indoor environment to be protected from high levels of VOCs includes any spaces where products and materials are applied on-site and within the weatherproofing system.

Industry standards, such as Green Seal (for paints, coatings, adhesives, etc.), South Coast Air Quality Management District (SCAQMD) (for sealants), the Carpet and Rug Institute (CRI), and the FloorScore™ Program specify the maximum levels of VOCs, in grams per liter (g/L), not to be exceeded. Other standards, such as GREENGUARD and Scientific Certification Systems (SCS) Indoor Advantage, certify products, such as furniture, that do not off-gas harmful levels of pollutants.

The SMACNA guidelines also address ventilation as a means to maintain quality indoor air, as well as the requirements for filtration media at all air returns (Figure 8.5). The LEED CI rating system follows suit with SMACNA practices and, therefore, also recommend employing **Minimum Efficiency Reporting Value (MERV) filters** to ensure effectiveness. The rating of MERV filters ranges from 1 (lowest) to 16 (highest), where LEED requires a minimum of MERV 8 filters to be implemented at return air intake locations for compliance with one credit, while requiring MERV 13 filters in another.

Create flashcard to remember MERV and 1–16 range of filters.

Figure 8.3 SMACNA-compliant practices include sealing off ductwork from dust and particulates.

INDOOR AIR QUALITY 191

Figure 8.4 Specifying materials with low to no VOCs helps to maintain good indoor air quality.
Photo courtesy of Sherwin-Williams

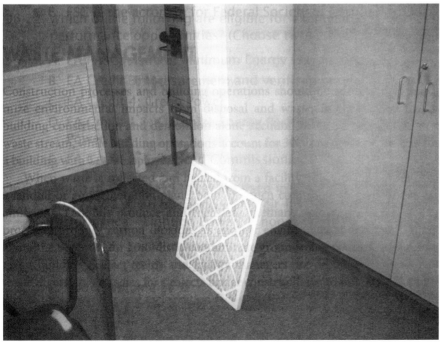

Figure 8.5 Installing MERV 8 filters or better will help to improve the indoor air quality by eliminating dust and particles.
Photo courtesy of Auld & White Constructors, LLC

Create a flashcard to remember the 5 SMACNA standards: HVAC protection, source control, pathway interruption, housekeeping, and scheduling.

The SMACNA guidelines also recommend good housekeeping practices during construction, to protect absorptive materials from moisture damage to later prevent the growth of toxic substances, such as mold. These materials, such as drywall, acoustic ceiling tiles, and carpet, should be stored in dry, elevated, and protected areas to avoid coming into contact with liquids (Figure 8.6).

EQ Credit 3.1: Construction IAQ Management Plan, During Construction. To comply, contractors will need to develop a Construction IAQ Management Plan to detail how the construction team will address the SMACNA guidelines during construction. The plan must include the strategies planned for the five aspects of SMACNA: heating, ventilation, and air conditioning (HVAC) protection; source control; pathway interruption; housekeeping; and scheduling. The plan will also need to require any permanently installed air handlers that are used during the construction phase to have a minimum of a MERV 8 filter at each return air grille to protect the ductwork. The filtration media must be replaced prior to occupancy.

EQ Credit 3.2: Construction IAQ Management Plan, Before Occupancy. LEED CI project teams have the opportunity to decide which of the two compliance paths to pursue for this credit. Typically, the decision is made because of limited time available in the schedule or the weather conditions at the time of year construction is completed. The first compliance path requires air quality testing in compliance with the EPA *Compendium of Methods for the Determination of Air Pollutants* in indoor air prior to occupancy to ensure nonharmful air levels for occupants. At least one sample must be taken for every 25,000 square feet. The reference guide lists the maximum concentration levels for contaminants, such as formaldehyde, particulates, total VOCs, 4-phenylcyclohexene (4-PCH), and carbon monoxide (CO) for the air quality testing to detect and determine compliance. If these contaminant levels are exceeded, the project has

Figure 8.6 Elevating product storage protects against damage as suggested by SMACNA guidelines.

failed the test, the building must be flushed out, and then the air retested in order to prove compliance.

The other compliance path option requires conducting a building flush-out to eliminate any pollutants in the air caused by construction processes and activities. This compliance path requires the mechanical system to be flushed out with 14,000 cubic feet of outside air per square foot to remove residual contaminants prior to occupancy. The key to complying is to maintain an internal temperature of at least 60°F and for the relative humidity not to exceed 60 percent within the building. If the schedule does not permit project teams to complete the latter or if occupancy is desired, they could also comply if at least 3,500 cubic feet of outside air per square foot is delivered prior to occupancy and then once the space is occupied, ventilation is started three hours prior to occupancy each day and is continued during occupancy at a rate of 0.30 cubic feet per minute (cfm) per square foot until the 14,000 cubic feet of air has been delivered.

Low-Emitting Materials

When addressing the following credits, a couple of characteristics and commonalities will be apparent. First, in order to prove compliance, project teams are required to track and list all applicable products for each credit. The list would indicate that all products used are in compliance with the coordinating reference standard. Should a member of the construction team disregard the requirements and use a product out of compliance, the project team has the opportunity to pursue the credit with the VOC budget methodology. This option demonstrates that the overall VOC performance was ascertained. Second, some of the reference standards overlap from one credit to the other so be sure to look out for these commonalities to help with study efforts.

The VOC Budget Method for EQ Credit 4.1 and 4.2

If an unapproved and/or noncompliant product is used within the weather barrier, it does not necessarily mean the team will not be able to pursue and earn the associated EQ low-emitting materials credit. In this case, the team will need to pursue the VOC budget method. With this method, a comparison is conducted for the total VOCs used on site (in g/L) to a baseline total maximum allowable VOC levels to document an overall low-VOC performance. For each product used at the project site, the total volume amount of product used is multiplied by the maximum VOC limit and then again by the actual VOC content of the product. This results in a total allowable VOC content and total VOC content used. As long as the total used is less than the total allowed, the project is in compliance. This comparison is completed separately for applicable products per credit. Therefore, if a noncompliant paint product is used, sealants and adhesives are not included in the calculations, as they are part of a different credit. Table 8.2 provides a sample approach for EQ Credit 4.2: Low-Emitting Materials, Paints and Coatings.

EQ Credit 4.1: Low-Emitting Materials, Adhesives and Sealants. All adhesives, sealants, and sealant primers used within the building must comply with SCAQMD Rule 1168, while aerosol adhesives must comply with the Green Seal Standard for Commercial Adhesives, GS-36.

TIP The Green Seal standards contain either GS or GC prefixes.

EQ Credit 4.2: Low-Emitting Materials, Paints and Coatings. All paints and coatings must comply with the Green Seal Standard GS-11. Anticorrosive/

Table 8.2 Sample VOC Budget Method Calculation for EQ Credit 4.2

Product	Volume of Product Used	Maximum VOC Allowed	VOC Budget	Actual VOC Content of Product	VOC Used
Interior Primer	150 liters	50 g/L	7,500	91 g/L	4,550
Eggshell Paint	50 liters	150 g/L	7,500	0 g/L	0
Product	Volume of Product Used	Maximum VOC Allowed	VOC Budget	Actual VOC Content of Product	VOC Used
Flat Paint	50 liters	50 g/L	2,500	97 g/L	48,50
Semigloss Paint	50 liters	150 g/L	7,500	0 g/L	0
Clear Wood Lacquer	38 liters	550 g/L	20,900	258/gL	9,804
TOTALS			45,900		19,204
					COMPLIANT

antirust paints must comply with the Green Seal Standard GC-3, while wood finishes, floor coatings, stains, primers, and shellacs used inside the building must comply with SCAQMD Rule 1113.

EQ Credit 4.3: Low-Emitting Materials, Flooring Systems. In order to meet the intentions of this credit, all carpet must comply with the Carpet and Rug Institute Green Label Plus program, and the carpet cushion must comply with the Carpet and Rug Institute Green Label program. Carpet adhesives must meet the requirements described in EQ Credit 4.1. The FloorScore™ Program standard is the compliance measure for all hard-surface flooring. Concrete, wood, bamboo, and cork floor finishes, such as stains and sealers, must be in compliance with EQ Credit 4.2 and meet SCAQMD Rule 1113. Tile-setting adhesives and grout must meet the requirements of EQ Credit 4.1 and comply with SCAQMD Rule 1168.

EQ Credit 4.4: Low-Emitting Materials, Composite Wood and Agrifiber Products. Composite wood and agrifiber products may also off-gas and, therefore, must be evaluated for compliance with this credit. The *ID+C Reference Guide* defines composite wood and agrifiber products as "particleboard, medium-density fiberboard (MDF), plywood, wheatboard, strawboard, panel substrates and door cores."[6] These products, including oriented-strand board (OSB), are manufactured with resin products to bind the fibers to hold the products together and form the useful building material. To comply within the parameters of this credit, CI projects with composite wood and agrifiber products cannot contain any added **urea-formaldehyde** resins, which off-gas at room temperature, to avoid contaminating the interior environment and, therefore, avoid causing health problems (Figure 8.7). Although many exterior building products may contain **phenol-formaldehyde,** this resin off-gases only at high temperatures and is, therefore, allowed within the confines of LEED compliance.

EQ Credit 4.5: Low-Emitting Materials, Systems Furniture and Seating. Project teams have three compliance path options from which to choose when pursuing this credit addressing all systems furniture and task and guest chairs

TIP Hard-surface flooring included in the FloorScore Program addresses linoleum, ceramic flooring, laminate flooring, vinyl, wood flooring, rubber flooring, and wall base.

 Notice the commonalities between SCAQMD Rules 1113 and 1168. On a similar note, which two credits share the requirements for carpet adhesive?

 Create a flashcard to remember that urea-formaldehyde is not allowed, but phenol-formaldehyde is suitable.

INDOOR AIR QUALITY 195

Figure 8.7 The Reston Association project in Reston, Virginia, by FOX Architects incorporated no added urea-formaldehyde wall finishes to help meet the credit requirements of EQ Credit 4.4, which helped the team to earn a Silver level of certification for the project. *Photo courtesy of John Cole Photography*

for the tenant occupants. Work tools that are attached to the panel-based or freestanding workstation do not have to comply with this credit in order for the project to pursue compliance. Furniture manufactured or refurbished within the year prior to occupancy are required to comply with one of the three compliance

Figure 8.8 Specifying GREENGUARD-certified furniture helps to maintain quality indoor air. *Photo courtesy of Steelcase, Inc.*

Create a flashcard to remember what VOCs are, that they are measured in grams per liter, and the different referenced standards associated with the different low-emitting material credits.

Create a flashcard to remember the seven credits to pursue during construction to improve IAQ.

paths. The first option is to specify and install all GREENGUARD Indoor Air Quality Certified products (Figure 8.8). The second option is to specify and install furniture and seating that has passed the EPA Environmental Technology Verification (ETV) Large Chamber Test Protocol for Measuring Emissions of VOCs and Aldehydes test. The third and final option requires the furniture and seating to pass the ANSI/BIFMA M7.1 2007 and ANSI/BIFMA x7.1-2007 testing protocols.

QUIZ TIME!

Q8.6. MERV filters range from which of the following? (Choose one)

    A. 1–30

    B. 0–50

    C. 1–110

    D. 1–16

    E. 1–100

Q8.7. Which of the following are the labeling standards applicable to carpets and carpet pads? (Choose two)

    A. CRI Green Label Plus

    B. Green-e

    C. Green carpets

    D. Green Seal

    E. Green Seal 36

    F. CRI Green Label

Q8.8. Which of the following are consistent with the requirements of EQ Credit 4.1: Low-Emitting Materials—Adhesives and Sealants? (Choose three)

    A. All adhesives and sealants must not exceed the VOC limits set by ASHRAE 232-1998: Maximum VOC Emissions in Occupied Spaces with Recirculating Air.

    B. Nonaerosol adhesives and sealants must be in compliance with VOC limits set by the SCAQMD Rule #1168.

    C. Paints and coatings must contain no phenol-formaldehyde.

    D. Adhesives and sealants must carry a Green Spec seal of approval.

    E. Aerosol adhesives must meet VOC limits established by the Green Seal Standard (GS-36).

    F. While projects are encouraged, but not required, to use low-VOC adhesives and sealants on exterior building elements, all adhesives and sealants inside of the building envelope weather seal must meet the requirements of the referenced standards.

Q8.9. SCAQMD Rule #1168 refers to which of the following? (Choose one)

   A. Measures air-change effectiveness
   B. Describes ventilation requirements for acceptable indoor air quality
   C. Defines the use of urea- and phenol-formaldehyde in composite wood and agrifiber products
   D. Sets the VOC maximum content for adhesives and sealants
   E. Sets the VOC maximum content for paints and coatings
   F. Defines the maximum VOC emissions from carpets and carpet cushions

Q8.10. Which of the following is not required by SMACNA standards? (Choose one)

   A. Source control
   B. Housekeeping
   C. HVAC protection
   D. Flush-out
   E. Scheduling

Q8.11. Which of the following components must comply in order for a project team to earn EQ Credit 4.5: Low-Emitting Materials—Systems Furniture and Seating? (Choose three)

   A. Private office desks
   B. Break room chairs
   C. Reception seating
   D. Freestanding workstations in an open office setting
   E. Task chairs
   F. Guest seating at a panel-based workstation

Q8.12. Which of the following are reference standards for EQ Credit 4.5: Low-Emitting Materials—Systems Furniture and Seating? (Choose three)

   A. GREENGUARD
   B. SCAQMD Rule #1168
   C. ANSI/BIFMA M7.1 2007
   D. Green Seal Standard (GS-36)
   E. EPA Environmental Technology Verification (ETV) Large Chamber Test Protocol for Measuring Emissions of VOCs and Aldehydes test

Prevention and Segregation Methods in Relation to LEED Compliance

Assessing and managing indoor pollutants has become a major concern in the battle against sick building syndrome. If buildings can be designed with the ability of maintaining a safe and healthy environment, building occupants will be more productive and have a greater well-being. The following strategies address means to reduce contamination for a cleaner indoor environment.

EQ Prerequisite 1: Environmental Tobacco Smoke (ETS) Control. Project teams have a few options from which to choose in order to comply with this prerequisite, depending on the use of the project. Compliance with option 1 requires the tenant space to be smoke free and be located in a building in which smoking is prohibited in all common areas and base building areas that share the same ventilation system. If smoking is allowed on-site, an area must be designated that is at least 25 feet away from any building entrances, outdoor air intakes, and operable windows. Option 2 allows smoking within the building, but only in designated areas. These areas must be separately exhausted to the outdoors so as not to contaminate other areas in the building. This would require mechanical engineers to be mindful of those exhaust areas not to be located next to intake zones or building entrances. The rooms should be enclosed with impermeable deck-to-deck partitions. Just as with option 1, if smoking will be allowed on-site, an area must be designated that is at least 25 feet away from any building entrances, outdoor air intakes, and operable windows.

Multi-unit residential building projects must prohibit smoking in all common areas within the building. All exterior doors and operable windows must be weatherstripped. All penetrations between units and vertical chases must be sealed to minimize contamination. Blower door tests must be conducted in accordance with ANSI/ASTM E779-03, Standard Test Method for Determining Air Leakage Rate by Fan Pressurization for residential units to prove successful sealing. Continuing with the requirements mentioned for options 1 and 2, where smoking will be allowed outdoors, an area must be designated that is at least 25 feet away from any building entrances, outdoor air intakes, and operable windows.

EQ Credit 5: Indoor Chemical and Pollutant Source Control. Project teams will need to comply with four strategies in order to earn this credit. First, all regularly used entrances must employ an entryway system that is at least 10 feet long in the primary direction of travel to reduce the amount of dirt and contaminants from entering the building. Second, the team must ensure that any areas where hazardous gases or chemicals will be present are separately exhausted, are provided with self-closing doors, and are enclosed with deck-to-deck partitions or a hard-lid ceiling (not to contaminate adjacent spaces). These areas could include high-volume copy areas, laundry areas, laboratories, and art rooms. The mechanical engineer needs to ensure that the HVAC system will accommodate a MERV 13 filter or better. Finally, containment must be provided for any hazardous materials to be properly disposed of and separate drains at any locations where the mixing of hazardous materials occurs. It is important to remember this credit requires all four components to be adhered to, not just one or a combination of strategies in order to earn this credit.

Remember, EQ Credit 3.1: Construction IAQ Management Plan requires a minimum of a MERV 8 filter during construction, whereas this credit requires a MERV 13 filter must be installed after construction but prior to occupancy. If needed, make a flashcard to remember the minimum MERV filters required and for which credit.

Create a flashcard to remember the four strategies to comply with EQ Credit 5.

Create a flashcard to remember the two strategies to improve IAQ by means of prevention and segregation.

Create a flashcard to remember the IAQ strategies include ventilation, construction practices, such as the specification and use of low-emitting materials, and prevention and segregation.

QUIZ TIME!

Q8.13. Which of the following are standards for indoor air quality? (Choose three)

A. GREENGUARD

B. ASHRAE 90.1

C. CRI Green Label Plus

D. Green Seal

E. ENERGY STAR

Q8.14. At areas of where hazardous gases or chemicals will be present, what are the requirements to comply with EQ Credit 5: Indoor Chemical and Pollutant Source Control? (Choose three)

   A. Separately exhausted
   B. Shared ventilation with adjacent spaces
   C. Self-closing doors
   D. Deck-to-deck partitions or a hard-lid ceiling
   E. Ceiling height walls

Q8.15. Which of the following are not required to comply with EQ Credit 5: Indoor Chemical and Pollutant Source Control? (Choose two)

   A. MERV 18
   B. MERV 13
   C. Disposal containers and drains for hazardous waste
   D. 10-foot-long entryway systems
   E. 6-foot-long entryway systems

Q8.16. If smoking is going to be allowed on-site, what is the minimum distance to designate smoking areas from building entrances, air intakes, and any operable windows in order to comply with EQ Prerequisite 2: ETS Control? (Choose one)

   A. 10 feet
   B. 15 feet
   C. 20 feet
   D. 25 feet
   E. 50 feet

THERMAL COMFORT

Although temperature settings should vary with the seasons, building owners are encouraged to allow for occupants to control their thermal conditions to optimize satisfaction and comfort. Remember, occupants who are satisfied and comfortable tend to be more productive! The *Green Building and LEED Core Concepts Guide* defines **thermal comfort** as "the temperature, humidity, and airflow ranges within which the majority of people are most comfortable, as determined by **ASHRAE Standard 55-2004**."[7] ASHRAE 55 indicates that there are four environmental factors that impact thermal comfort determined by the building design: humidity, air speed, air temperature, and radiant temperature. For the purposes of LEED, occupants must be able to control one of the four components of thermal comfort in order to comply with the control strategy.

Anytime you see ASHRAE 55, think THERMAL COMFORT. What do you think of when you see ASHRAE 90.1? How about ASHRAE 62?

Create a flashcard to remember the four environmental factors of thermal comfort defined by ASHRAE 55.

Thermal Comfort in Relation to LEED Compliance

EQ Credit 6.2: Controllability of Systems, Thermal Comfort. Project teams will need to provide access to control at least one of the four factors of thermal comfort for at least 50 percent of the building occupants in regularly occupied areas to comply with this credit. Shared multioccupant areas, such as conference rooms, classrooms, or lecture halls, must have at least one method to control the thermal environment. Operable windows may be used for compliance instead of controls (Figure 8.9) if the occupants are located within 20 feet inside and no more than 10 feet side to side of the windows, and the windows comply with the requirements of ASHRAE 62.1-2007, paragraph 5.1, Natural Ventilation. Mechanically ventilated projects could implement a raised access floor for under-floor air distribution to comply with the intentions of this credit (Figure 8.10).

EQ Credit 7.1: Thermal Comfort, Design. The mechanical engineer will need to ensure that the HVAC system is designed in accordance with ASHRAE

Figure 8.9 Wausau's LEED Silver facility in Wausau, Wisconsin, employs operable windows to give their employees access to fresh air. *Photo courtesy of Wausau Windows and Wall Systems*

Figure 8.10 Raised access floors provide the flexibility to grant occupants individual control of the amount of air supplied through diffusers for their thermal comfort.
Photo courtesy of Tate Access Floors, Inc.

55-2004 in order to comply with the intentions of this credit. In addition to the four building design factors of thermal comfort mentioned previously, the engineer will also need to address two other factors that are determined by occupancy: metabolic rate and clothing insulation.

EQ Credit 7.2: Thermal Comfort, Verification. In addition to pursuing the previous credit, building owners will need to commit to conducting an anonymous occupant survey within 6 to 18 months after occupying the building to discover the overall satisfaction of the thermal comfort levels of the majority of the occupants to determine areas for improvement. A plan for corrective action must also be in place in order to comply with this credit. In order to ensure that building performance is in accordance with ASHRAE 55-2004, projects must also include a permanent monitoring system.

 Create a flashcard to remember the three credits that address thermal comfort.

QUIZ TIME!

Q8.17. When addressing thermal comfort, which two of the following are not addressed? (Choose two)

    A. Humidity

    B. Ventilation requirements

    C. Air movement

    D. Artificial light

    E. Average temperature

CHAPTER 8: INDOOR ENVIRONMENTAL QUALITY

Q8.18. An Environmental Tobacco Smoke Control policy *best* addresses which of the following? (Choose one)

A. Providing ventilation requirements to effectively remove tobacco smoke

B. Providing dedicated smoking rooms 25 feet away from building entrances

C. Preventing tobacco smoke from contaminating indoor environments

D. Preventing tobacco smoke from entering the air occupied by nonsmokers

Q8.19. How long after occupancy should a thermal comfort survey be conducted to be in accordance with EQ Credit 7.2: Thermal Comfort, Verification? (Choose one)

A. Within 18 months

B. Within 12 months

C. Between 6 and 18 months

D. Between 6 and 12 months

E. Within 6 months

Q8.20. What is the minimum percentage of occupants that must have thermal control in order to comply with EQ Credit 7: Controllability of Systems, Thermal Comfort? (Choose one)

A. 10 percent

B. 25 percent

C. 90 percent

D. 50 percent

E. 75 percent

LIGHTING

The LEED CI rating system addresses lighting in terms of naturally available daylight and artificially supplied light. When debating whether to incorporate daylighting strategies, project teams are advised to conduct a life-cycle cost analysis to determine the up-front costs and operational savings. For example, when using daylighting strategies, sensors could be installed to trigger alternative light sources when needed, which would impact up-front costs, although the costs can be offset by the energy saved during operations since less artificial light would be required. Daylighting can also result in improved occupant satisfaction and health because of access and connection to the exterior environment, also affecting the economic bottom line over time (Figure 8.11).

Besides daylighting, providing occupants with consistent, ergonomic, and the ability to control their lighting needs can also benefit the triple bottom line. For example, occupant-controlled lighting contributes to employee satisfaction, as well as productivity, as light levels can be altered for specific tasks, needs, and preferences (Figure 8.12). Therefore, providing overall ambient light, as well as individual task lighting, is the best strategy to address lighting needs. Facilities

Figure 8.11 Providing interior environments with access to natural daylight not only improves the occupant's satisfaction and productivity levels but also helps to reduce the need for artificial lighting to reduce operating costs.
Photo courtesy of Steelcase, Inc.

Figure 8.12 Providing occupants with task lighting allows for more individual control of work environments to improve their satisfaction and productivity.
Photo courtesy of FXFOWLE Architects and Eric Laignel

can also see a reduction in energy usage for lighting needs by educating employees on the benefits of turning off fixtures after use.

Lighting in Relation to LEED Compliance

The LEED CI rating system offers the following three strategies to address lighting for projects seeking LEED certification:

EQ Credit 6.1: Controllability of Systems, Lighting. CI project teams will need to provide lighting controls for at least 90 percent of the tenant occupants and all shared **group multioccupant spaces**, such as conference rooms and classrooms.

EQ Credit 8.1: Daylight and Views, Daylight. For CI projects, teams must provide at least 75 percent of regularly occupied areas with daylight illumination to comply. Projects that achieve 95 percent daylighting are eligible to pursue an Exemplary Performance credit.

The most important factor for teams to address is consistent across all compliance path options for this credit: glare control (Figure 8.13). This strategy is important to address because avoiding it can hinder the productivity and comfort of the building occupants. Solar heat gain is another factor for teams to consider and address as it directly impacts energy efficiency. Different types of glass perform in different ways, depending on how much light is reflected, absorbed, and transmitted. Transmitted light, or T_{vis}, and the amount of light absorbed affects solar heat gain. For daylighting purposes, it is best to select glazing with a high T_{vis} to allow for the most amount of incident light to pass through the glazing. Therefore, project teams are challenged to find the balance between solar heat gain and T_{vis}.

Project teams can choose whether to pursue a performance-based or a prescriptive-based compliance path, similarly to pursuing the energy performance prerequisite and credit within the EA category. The performance-based option includes a computer simulation demonstrating footcandle (fc) levels between 25 and 500 provided by daylight for the applicable spaces. The prescriptive path involves completing calculations to show compliance depending on top-lighting daylighting zones (using skylights) and side-lighting daylighting zones (using

> When one is designing, non–regularly occupied spaces should be placed at the core to allow offices, classrooms, and other regularly occupied areas to be placed along the window line.

> Remember, building orientation and passive design strategies affect the opportunity to use daylighting as an ambient light source. There is a direct correlation between the amount of vision glazing provided and the access to daylight and views.

Figure 8.13 The lighting design at The National Audubon Society project in New York designed by Illumination Arts and FXFOWLE Architects includes simple downlights and wallwashers in the conference rooms where both were separately controlled by dimmers, thus allowing the occupants to adjust the lighting to fit their needs and the task at hand. Notice too the glare control strategy at the window. These types of strategies helped to earn the project its LEED CI Platinum level certification. *Photo courtesy of FXFOWLE Architects and David Sundberg/Esto*

windows or glass doors). Top-lighting daylight strategies must include skylights with a minimum 0.5 visible light transmittance (VLT) for at least 3–6 percent of the roof area, and the distance between the skylights must not be more than 1.4 times the ceiling height. Only glazing areas 30 inches above the floor can be included in side-lighting daylight calculations. In this zone, the window-to-floor ratio (WFR) multiplied by the VLT must be between 0.150 and 0.180 in order to comply with the requirements of this option. A third option is available for teams wishing to wait until the space is constructed in order to measure the actual light levels in the applicable spaces and a fourth option can be pursued to show compliance with this credit by combining of any of the three other options.

EQ Credit 8.2: Daylight and Views, Views for Seated Spaces. Project teams will need to provide 90 percent of the regularly occupied areas with views to the exterior by the means of vision glazing in order to meet the requirements of this credit. Remember, vision glazing is the glass from 30 inches to 90 inches above the floor. Project teams will need to upload a plan view to show all of the perimeter glazing and a section view to depict the line of sight from the regularly occupied area to the perimeter glazing with the line of sight at 42 inches above the floor. Exemplary performance opportunities are available for projects that meet 2 of the 4 listed criteria in the *ID+C Reference Guide*. The criteria addresses multiple lines of sight, type of view, unobstructed views, and view factor.

TIP Project teams should design the overall illumination light level to be about 40–60 footcandles on an office work surface.

TIP For private offices, all of the square footage may count toward EQ credit 8.2 compliance if 75 percent or more has a view to the exterior. This concept cannot be applied to multioccupant areas, as only the actual square footage that is compliant can be included in the calculations to achieve the minimum percentage threshold.

Create a flashcard to remember the three EQ credits that address lighting.

QUIZ TIME!

Q8.21. What is the most common failure of daylighting strategies? (Choose one)

   A. Shallow floorplates
   B. Interior color schemes
   C. Shading devices
   D. Glare control

Q8.22. Which of the following are strategies to control glare? (Choose three)

   A. Blinds
   B. Louvers
   C. Light shelves
   D. Reflective glazing
   E. Building orientation
   F. Lower ceiling heights

Q8.23. In addition to addressing light levels, what are other strategies are to be considered when designing for daylighting? (Choose three)

   A. Photovoltaic system energy contribution
   B. Direct beam penetration
   C. Integration with the electric lighting system
   D. Mechanical heat and cooling
   E. Interior color schemes

Q8.24. At what height is the line of sight drawn in section view in order to comply with EQ Credit 8.2: Daylight and Views, Views for Seated Spaces? (Choose one)

A. 30 inches

B. 32 inches

C. 34 inches

D. 36 inches

E. 42 inches

Q8.25. Which one of the following most closely represents an appropriate level of overall illumination on an office work surface, including daylighting, ambient artificial lighting, and task lighting? (Choose one)

A. 1–2 footcandles

B. 5–10 footcandles

C. 15–25 footcandles

D. 40–60 footcandles

E. 75–120 footcandles

F. 150–200 footcandles

Q8.26. A tenant is going to lease 60,000 square feet in an existing timber frame building. This space includes exposed timber beams, but the tenant plans on replacing the single-pane windows with energy-efficient glazing manufactured in a nearby town. Salvaged wood floors will be donated from an adjacent property. Which of the following strategies can the project team pursue? (Choose three)

A. MR Credit 7: Certified Wood

B. EQ Credit 4.4: Low-Emitting Materials, Composite Wood and Agrifiber Products

C. EA Prerequisite 2: Minimum Energy Performance

D. MR Credit 5: Regional Materials

E. MR Credit 3.1: Materials Reuse

Q8.27. Which of the following best describes the LEED strategy applicable to ASHRAE Standard 62.1-2007? (Choose one)

A. Exterior lighting levels

B. Thermal comfort by the means of controllability of systems

C. Environmental Tobacco Smoke (ETS) Control

D. Ventilation and indoor air quality

E. Building flush-out parameters and guidelines

Q8.28. A project team plans to use a raised access floor to include under-floor air distribution, allowing the use of floor-mounted operable diffusers at each workstation, thereby eliminating overhead ducts and maximizing the interior floor-to-ceiling height. The moderate supply air temperature required at the diffusers would reduce the amount of energy associated with cooling a consistently greater quantity of outside air needed to improve air quality. Which two LEED strategies are addressed? (Choose two)

   A. Increased Ventilation
   B. Construction IAQ Management
   C. Regional Priority
   D. Controllability of Systems: Thermal Comfort
   E. Acoustical Performance

Q8.29. Which of the following design team deliverables and team members are most likely to play a significant role in achieving Construction IAQ Management Plan? (Choose two)

   A. Project specifications
   B. Civil engineer
   C. Construction documents
   D. Lighting designer
   E. General contractor
   F. Electrical engineer

Q8.30. To which standards should engineers design the ventilation systems for a LEED project? (Choose two)

   A. ASHRAE 55
   B. California Air Resources Board
   C. ASHRAE 62.1
   D. ASHRAE 90.1
   E. ASTM 44

Q8.31. Which of the following strategies have proven to increase productivity and occupant satisfaction in green buildings? (Choose two)

   A. Providing access to daylight
   B. Selecting a site adjacent to a shopping center
   C. Improving indoor air quality
   D. Implementing a recycling program
   E. Offering incentives for carpooling

Q8.32. Which of the following EQ credits are eligible to be submitted after substantial completion for certification review? (Choose two)

A. Credit 4: Low-Emitting Materials
B. Credit 2: Increased Ventilation
C. Credit 3: Construction IAQ Management Plan
D. Credit 8: Daylight and Views
E. Credit 1: Outdoor Air Delivery Monitoring

Q8.33. Which of the following are examples of a regularly occupied space? (Choose three)

A. Laundry room
B. Conference room
C. Restroom
D. Private office
E. Cafeteria
F. Storage room

Q8.34. Which of the following are examples of a non–regularly occupied space? (Choose three)

A. Laundry room
B. Conference room
C. Restroom
D. Private office
E. Storage room

Q8.35. Which of the following is not related to EQ Credit 2: Increased Ventilation? (Choose one)

A. EA Prerequisite 1 and Credit 3: Fundamental and Enhanced Commissioning
B. EA Prerequisite 2 and Credit 1: Minimum and Enhanced Energy Performance
C. EQ Credit 7: Thermal Comfort
D. EA Credit 5: Measurement and Verification
E. EQ Credit 1: Outdoor Air Delivery Monitoring

Q8.36. A 22,300 square foot tenant improvement project is seeking EQ Credit 3.2: Construction IAQ Management, Before Occupancy. How much outside air will be required for a flush-out to earn the point? (Choose one)

A. 14,000 cu ft
B. 312,200,000 cu ft
C. 78,050,000 cu ft
D. 234,150,000 cu ft
E. There is not enough information provided

DESIGN	**EQ Prerequisite 1:** **Minimum Indoor Air Quality Performance**
	EQ Credit 2: **Increased Ventilation**

PURPOSE

Prerequisite:

To improve the indoor air quality in buildings to enhance the _____ and _____ - _____ of the occupants by establishing a _____ indoor air quality performance.

Credit:

To increase the _____ _____ _____ in buildings to enhance the comfort, well-being, and _____ of the occupants by providing _____ outdoor air ventilation.

REQUIREMENTS

Case 1: Mechanically Ventilated Buildings

Prerequisite:

Comply with either the _____ code or ASHRAE 62.1 - 2007 (whichever requires more outside air). If the project _____ meet the minimum levels of ASHRAE 62.1, confirm that an absolute minimum of 10 cubic feet per minute per person is supplied to the space and _____ why the project is not able to meet the requirements of the referenced standard.

Credit:

Provide at least _____ percent more outdoor air than that required by the Prerequisite.

Case 2: Naturally Ventilated Buildings

Prerequisite:

Comply with ASHRAE 62.1 - 2007, Paragraph 5.1. If the project cannot meet the minimum levels of ASHRAE 62.1, confirm that an absolute minimum of _____ cubic feet per _____ per _____ is supplied to the space and document why the project is not able to meet the requirements of the referenced standard.

Credit: Meet recommendations of _____ Trust's *Good Practice Guide* _____ for occupied spaces and one of the following options:

Option 1: Design building to meet recommendations described in _____ *Applications Manual 10*

Option 2: Predict _____ -by- _____ airflows with an analytic model to meet the minimum ventilation rates required by ASHRAE 62.1 - 2007 for at least _____ percent of the occupied spaces.

REFERENCED STANDARDS

ASHRAE Standard 62.1 - 2007, _____ for Acceptable Indoor Air Quality

_____ *Applications Manual 10 - 2005, Natural Ventilation in Non-Domestic Buildings*

The Carbon Trust's *Good Practice Guide 237, Natural Ventilation in Non-Domestic Building, A Guide for Designers, Developers, and Owners*

5 pts for credit	RESPONSIBLE PARTY: **MECHANICAL ENGINEER**

EQ Prerequisite 1:
Minimum Indoor Air Quality Performance

EQ Credit 2:
Increased Ventilation

ANSWER KEY

PURPOSE

Prerequisite:

To improve the indoor air quality in buildings to enhance the **comfort** and **well-being** of the occupants by establishing a **minimum** indoor air quality performance.

Credit:

To increase the **indoor air quality** in buildings to enhance the comfort, well-being, and **productivity** of the occupants by providing **additional** outdoor air ventilation.

REQUIREMENTS

Case 1: Mechanically Ventilated Buildings

Prerequisite:

Comply with either the **local** code or ASHRAE 62.1 - 2007 (whichever requires more outside air). If the project **cannot** meet the minimum levels of ASHRAE 62.1, confirm that an absolute minimum of 10 cubic feet per minute per person is supplied to the space and **document** why the project is not able to meet the requirements of the referenced standard.

Credit:

Provide at least **30** percent more outdoor air than that required by the Prerequisite.

Case 2: Naturally Ventilated Buildings

Prerequisite:

Comply with ASHRAE 62.1 - 2007, Paragraph 5.1. If the project cannot meet the minimum levels of ASHRAE 62.1, confirm that an absolute minimum of **10** cubic feet per **minute** per **person** is supplied to the space and document why the project is not able to meet the requirements of the referenced standard.

Credit: Meet recommendations of **Carbon** Trust Good Practice Guide **237** for occupied spaces and one of the following options:

Option 1: Design building to meet recommendations described in **CIBSE** Applications Manual 10

Option 2: Predict **room**-by-**room** airflows with an analytic model to meet the minimum ventilation rates required by ASHRAE 62.1 - **2007** for at least **90** percent of the occupied spaces.

REFERENCED STANDARDS

ASHRAE Standard 62.1 - 2007, **Ventilation** for Acceptable Indoor Air Quality

CIBSE Applications Manual 10 - 2005, Natural Ventilation in Non-Domestic Buildings

The Carbon Trust's Good Practice Guide 237, Natural Ventilation in Non-Domestic Building, A Guide for Designers, Developers, and Owners

DESIGN

EQ Prerequisite 2:

Environmental Tobacco Smoke (ETS) Control

PURPOSE

To _____ or minimize environmental _____ smoke (ETS) from contaminating the indoor environment, including occupants, _____, and ventilation systems.

REQUIREMENTS

Option 1:

    Do not allow smoking in the building.

    Provide a dedicated smoking area on site that is at least 25 feet away from any building entrances, outdoor air _____ and operable windows.

Option 2:

    Do not allow smoking within the _____ areas (and any areas that share the same _____ system), except in dedicated areas with separate and isolated ventilation and exhaust systems. Interior smoking areas also to have impermeable deck-to-deck partitions.

    Provide a dedicated smoking area on site that is at least _____ feet away from any building entrances, outdoor _____ intakes and _____ windows.

Option 3: Residential and Hospitality Projects Only

    Do not allow smoking in the common areas of the building, tenant space, and shared ventilation spaces with the tenant.

    Provide a _____ exterior smoking area on site that is at least _____ feet away from any building entrances, outdoor air intakes and operable windows (this includes balconies).

    _____ all openings and seal all penetrations in each unit.

    Conduct a _____ door test in accordance with reference standard to demonstrate acceptable _____ of units.

DOCUMENTATION

Smoking policy and designated smoking areas shown on a plan

REFERENCED STANDARDS

ANSI/ASTM _____, Standard Test Method for Determining Air Leakage Rate by Fan Pressurization

Residential Manual for Compliance with California's 2001 Energy Efficiency Standards (for Low Rise Residential Buildings), Chapter 4

Required

RESPONSIBLE PARTY:

OWNER

SS Prerequisite 2:
Environmental Tobacco Smoke (ETS) Control

ANSWER KEY

PURPOSE

To **prevent** or minimize environmental **tobacco** smoke (ETS) from contaminating the indoor environment, including occupants, **surfaces**, and ventilation systems.

REQUIREMENTS

Option 1:

Do not allow smoking in the building.

Provide a dedicated smoking area on site that is at least 25 feet away from any building entrances, outdoor air **intakes** and operable windows.

Option 2:

Do not allow smoking within the **tenant** areas (and any areas that share the same **ventilation** system), except in dedicated areas with separate and isolated ventilation and exhaust systems. Interior smoking areas also to have impermeable deck-to-deck partitions.

Provide a dedicated smoking area on site that is at least **25** feet away from any building entrances, outdoor **air** intakes and **operable** windows.

Option 3: Residential and Hospitality Projects Only

Do not allow smoking in the common areas of the building, tenant space, and shared ventilation spaces with the tenant.

Provide a **dedicated** exterior smoking area on site that is at least **25** feet away from any building entrances, outdoor air intakes and operable windows (this includes balconies).

Weatherstrip all openings and seal all penetrations in each unit.

Conduct a **blower** door test in accordance with reference standard to demonstrate acceptable **sealing** of units.

DOCUMENTATION

Smoking policy and designated smoking areas shown on a plan

REFERENCED STANDARDS

ANSI/ASTM **E779-03**, Standard Test Method for Determining Air Leakage Rate by Fan Pressurization

Residential Manual for Compliance with California's 2001 Energy Efficiency Standards (for Low Rise Residential Buildings), Chapter 4.

DESIGN

EQ Credit 1:
Outdoor Air Delivery Monitoring

PURPOSE

Promote occupant comfort and well-being by providing ventilation system _____.

REQUIREMENTS

Provide _____ monitoring _____ sensors within the breathing zone (_____ to _____ feet above the floor) for all densely occupied spaces for mechanically ventilated buildings and all spaces within a naturally ventilated building. For _____ occupied spaces, install outdoor _____ measuring devices.

Sensors must provide an audible or visual alarm if CO_2 conditions vary by more than _____ percent.

Outdoor _____ rates must also be monitored to ensure ventilation effectiveness.

REFERENCED STANDARD

ASHRAE Standard _____ - 2007, Ventilation for _____ Indoor Air Quality

1 pt

RESPONSIBLE PARTY:
MECHANICAL ENGINEER

EQ Credit 1: Outdoor Air Delivery Monitoring

ANSWER KEY

PURPOSE

Promote occupant comfort and well-being by providing ventilation system **monitoring**.

REQUIREMENTS

Provide **permanent** monitoring CO_2 sensors within the breathing zone (**3** to **6** feet above the floor) for all densely occupied spaces for mechanically ventilated buildings and all spaces within a naturally ventilated building. For **nondensely** occupied spaces, install outdoor **intake** measuring devices.

Sensors must provide an audible or visual alarm if CO_2 conditions vary by more than **10** percent.

Outdoor **airflow** rates must also be monitored to ensure ventilation effectiveness.

REFERENCED STANDARD

ASHRAE Standard **62.1** - 2007, Ventilation for **Acceptable** Indoor Air Quality

CONSTRUCTION

EQ Credit 3.1:
Construction IAQ Management Plan,
During Construction

PURPOSE

Promote occupant _____ and well-being by reducing _____ air quality problems caused during _____.

REQUIREMENTS

Create and implement a _____ management plan during construction to comply with _____ IAQ Guidelines.

All _____ materials must be protected from _____ damage whether they are installed or stored on-site.

Use a minimum of MERV _____ filters at all _____ air grilles for any permanently installed air handlers are used during construction, per ASHRAE Standard _____. All filters to be replaced prior to occupancy.

STRATEGIES

_____ Guidelines:

Source Control

Pathway Interruption

_____

_____

HVAC Protection

REFERENCED STANDARDS

_____ Metal and Air Conditioning Contractors National Association (SMACNA) IAQ Guidelines for Occupied Buildings Under _____ 2nd Edition 2007, ANSI/SMACNA 008-2008 (Chapter 3)

ANSI/ASHRAE Standard 52.2 - 1999: Method of Testing General Ventilation Air-Cleaning Devices for Removal Efficiency by _____ Size

1 pt

RESPONSIBLE PARTY:
CONTRACTOR

EQ Credit 3.1:
Construction IAQ Management Plan, During Construction

ANSWER KEY

PURPOSE

Promote occupant **comfort** and well-being by reducing **indoor** air quality problems caused during **construction**.

REQUIREMENTS

Create and implement a **IAQ** management plan during construction to comply with **SMACNA** IAQ Guidelines.

All **absorptive** materials must be protected from **moisture** damage whether they are installed or stored on-site.

Use a minimum of MERV **8** filters at all **return** air grilles for any permanently installed air handlers are used during construction, per ASHRAE Standard **52.2**. All filters to be replaced prior to occupancy.

STRATEGIES

SMACNA Guidelines:

    Source Control

    Pathway Interruption

    Housekeeping

    Scheduling

    HVAC Protection

REFERENCED STANDARDS

Sheet *Metal and Air Conditioning Contractors National Association (SMACNA) IAQ Guidelines for Occupied Buildings Under Construction*, 2nd Edition 2007, ANSI/SMACNA 008-2008 (Chapter 3)

ANSI/ASHRAE Standard 52.2 - 1999: Method of Testing General Ventilation Air-*Cleaning* Devices for Removal Efficiency by **Particle** Size

	EQ Credit 3.2:
CONSTRUCTION	**Construction IAQ Management Plan, Before Occupancy**

PURPOSE

Promote occupant _____ and well-being by reducing _____ air quality problems caused during _____.

REQUIREMENTS

Option 1: Flush-out

Provide _____ cubic feet of outside air per square foot of floor area maintain an average of 60°F with a maximum of _____ percent relative humidity.

If occupancy is desired, provide at least _____ cubic feet of outside air per square foot of floor area. After occupancy, begin ventilation at least 3 hours prior to occupancy each day at a minimum rate of _____ cubic feet per minute per square foot of outside air (or the design minimum of EQ Prerequisite 1—whichever is greater) until _____ cubic feet of outside air per square foot of floor area has been delivered.

Option 2: Air Testing

Complete testing after _____ but prior to _____ in accordance with the protocols of the reference standard.

Samples must be collected during normal occupied hours and within _____ zone for at least _____ hours in areas with the _____ amount of ventilation.

If contaminant levels exceed threshold, the project may fail the test. If this is the case, the space will need to be _____ _____ and _____.

REFERENCED STANDARD

U.S. EPA _____ of Methods for the Determination of Air _____ in Indoor Air

1 pt

RESPONSIBLE PARTY: **CONTRACTOR**

EQ Credit 3.2: Construction IAQ Management Plan, Before Occupancy

ANSWER KEY

PURPOSE

Promote **occupant** comfort and well-being by reducing **indoor** air quality problems caused during **construction**.

REQUIREMENTS

Option 1: Flush-out

Provide **14,000** cubic feet of outside air per square foot of floor area maintain an average of 60°F with a maximum of **60** percent relative humidity.

If occupancy is desired, provide at least **3,500** cubic feet of outside air per square foot of floor area. After occupancy, begin ventilation at least **3** hours prior to occupancy each day at a minimum rate of **0.30** cubic feet per minute (cfm) per square foot of outside air (or the design minimum of EQ Prerequisite 1—whichever is greater) until **14,000** cubic feet of outside air per square foot of floor area has been delivered.

Option 2: Air Testing

Complete testing after **construction** but prior to **occupancy** in accordance with the protocols of the reference standard.

Samples must be collected during normal occupied hours and within **breathing** zone for at least **4** hours in areas with the **least** amount of ventilation.

If contaminant levels exceed threshold, the project may fail the test. If this is the case, the space will need to be **flushed out** and **retested**.

REFERENCED STANDARD

U.S. EPA **Compendium** of Methods for the Determination of Air **Pollutants** in Indoor Air

CONSTRUCTION

**EQ Credit 4.1:
Low-Emitting Materials,
Adhesives and Sealants**

PURPOSE

To limit the amount of _____, irritating and/or harmful contaminants used _____ to enhance the _____ and well-being of _____ and occupants.

REQUIREMENTS

_____, sealants, and sealant _____ must not exceed the VOC limits of SMACNA _____.

Aerosol adhesives must not exceed the VOC limits of Green Seal Standard 36.

DOCUMENTATION

Contractor to submit a schedule of all adhesives, sealants, and sealant primers used on the interior of the space (inside of the _____ system and applied on-site) with VOC level expressed in _____.

If not all products comply, contractor is to submit a _____ _____ to prove the _____ of all products comply.

REFERENCED STANDARDS

South _____ Air Quality _____ District (SCAQMD) Rule #1168

Green Seal Standard 36

1 pt

RESPONSIBLE PARTY:
CONTRACTOR

EQ Credit 4.1: Low-Emitting Materials, Adhesives and Sealants

ANSWER KEY

PURPOSE

To limit the amount of **odorous**, irritating and/or harmful contaminants used **indoors** to enhance the **comfort** and well-being of **installers** and occupants.

REQUIREMENTS

Adhesives, sealants, and sealant **primers** must not exceed the VOC limits of SMACNA **1168**.

Aerosol adhesives must not exceed the VOC limits of Green Seal Standard 36.

DOCUMENTATION

Contractor to submit a schedule of all adhesives, sealants, and sealant primers used on the interior of the space (inside of the **weatherproofing** system and applied on-site) with VOC level expressed in **g/L**.

If not all products comply, contractor is to submit a **VOC budget** to prove the **average** of all products comply.

REFERENCED STANDARDS

South **Coast** Air Quality **Management** District (SCAQMD) Rule #1168

Green Seal Standard 36

CONSTRUCTION

**EQ Credit 4.2:
Low-Emitting Materials,
Paints and Coatings**

PURPOSE

To limit the amount of _____, irritating and/or harmful contaminants used _____ to enhance the _____ and well-being of _____ and occupants.

REQUIREMENTS

_____ and coatings must not exceed the VOC limits of Green _____ Standard _____.

Anticorrosive and _____ paints must not exceed the VOC limits of _____ Seal Standard _____.

Clear wood finishes, _____ _____, stains, primers, and shellacs applied to interior surfaces must not exceed the VOC limits of SMACNA #_____.

DOCUMENTATION

Contractor to submit a schedule of all _____ and _____ used on the interior of the space (inside of the weatherproofing system and applied _____ _____) with _____ level expressed in g/L.

If not all products comply, contractor is to submit a VOC budget to prove the average of all products comply.

REFERENCED STANDARDS

South Coast Air Quality Management District (SCAQMD) Rule #1113

_____ _____ Standard 11 (GS-11)

Green Seal Standard _____ (_____ _____)

1 pt

RESPONSIBLE PARTY:
CONTRACTOR

EQ Credit 4.2: Low-Emitting Materials, Paints and Coatings

ANSWER KEY

PURPOSE

To limit the amount of **odorous**, irritating and/or harmful contaminants used **indoors** to **enhance** the comfort and well-being of **installers** and occupants.

REQUIREMENTS

Paints and coatings must not exceed the VOC limits of Green **Seal** Standard **GS-11**.

Anticorrosive and **antirust** paints must not exceed the VOC limits of **Green** Seal Standard **GC-3**.

Clear wood finishes, **floor coatings**, stains, primers, and shellacs applied to interior surfaces must not exceed the VOC limits of SMACNA #**1113**.

DOCUMENTATION

Contractor to submit a schedule of all **paints** and **coatings** used on the interior of the space (inside of the weatherproofing system and applied **on-site**) with **VOC** level expressed in g/L.

If not all products comply, contractor is to submit a VOC budget to prove the average of all products comply.

REFERENCED STANDARDS

South Coast Air Quality Management District (SCAQMD) Rule #1113

Green Seal Standard 11 (GS-11)

Green Seal Standard 3 (**GC-3**)

CONSTRUCTION

EQ Credit 4.3:
Low-Emitting Materials,
Flooring Systems

PURPOSE

To limit the amount of _____, irritating and/or harmful contaminants used _____ to enhance the _____ and well-being of _____ and occupants.

REQUIREMENTS

All _____ must comply with The Carpet and Rug Institute (CRI) Green Label _____ program.

All carpet _____ must comply with The Carpet and Rug Institute (CRI) _____ Label program.

All carpet adhesive must comply with the requirements of EQ Credit _____: Adhesives and Sealants.

All _____ surface flooring must comply with the requirements of the _____ standard and be certified by an independent third-party. Hard surface flooring includes vinyl, _____, laminate, wood, ceramic, rubber, and _____ base.

Concrete, _____, bamboo, and cork floor finishes (sealer, _____, and finish) must not exceed the VOC limits of SMACNA #_____.

Tile setting adhesives and grout must not exceed the VOC limits of SMACNA #_____.

OR

_____ flooring products must comply with the California Department of Health Services Standard Practice for the Testing of Volatile Organic _____ from Various Sources Using Small-Scale Environmental Chambers.

DOCUMENTATION

Contractor to submit a schedule of all flooring products, _____, grout, and finishes used on the project and documentation proving compliance, including _____ levels.

REFERENCED STANDARDS

The Carpet and _____ Institute(CRI) Green Label _____ Testing Program

South _____ Air Quality Management District (SCAQMD) Rule #1168

South Coast _____ Quality _____ District (SCAQMD) Rule #1113

_____ Program

_____ Department of _____ Services Standard Practice for the _____ of _____ Organic Emissions from Various Sources Using Small-Scale Environmental Chambers

1 pt

RESPONSIBLE PARTY:
CONTRACTOR

EQ Credit 4.3: Low-Emitting Materials, Flooring Systems

ANSWER KEY

PURPOSE

To limit the amount of **odorous**, irritating and/or harmful contaminants used **indoors** to enhance the **comfort** and well-being of **installers** and occupants.

REQUIREMENTS

All **carpet** must comply with The Carpet and Rug Institute Green Label **Plus** program.

All carpet **cushion** must comply with The Carpet and Rug Institute **Green** Label program.

All carpet adhesive must comply with the requirements of EQ Credit **4.1**: Adhesives and Sealants.

All hard surface flooring must comply with the requirements of the **FloorScore** standard and be certified by an independent third-party. Hard surface flooring includes vinyl, **linoleum**, laminate, wood, ceramic, rubber, and **wall** base.

Concrete, **wood**, bamboo, and cork floor finishes (sealer, **stain**, and finish) must not exceed the VOC limits of SMACNA #**1113**.

Tile setting adhesives and grout must not exceed the VOC limits of SMACNA #**1168**.

OR

All flooring products must comply with the California Department of Health Services Standard Practice for the Testing of Volatile Organic **Emissions** from Various Sources Using Small-Scale Environmental Chambers.

DOCUMENTATION

Contractor to submit a schedule of all flooring products, **adhesives**, grout, and finishes used on the project and documentation proving compliance, including **VOC** levels.

REFERENCED STANDARDS

The Carpet and **Rug** Institute (CRI) Green Label **Plus** Testing Program

South **Coast** Air Quality Management District (SCAQMD) Rule #1168

South Coast **Air** Quality **Management** District (SCAQMD) Rule #1113

FloorScore Program

California Department of **Health** Services Standard Practice for the **Testing** of **Volatile** Organic Emissions from Various Sources Using Small-Scale Environmental Chambers

CONSTRUCTION

EQ Credit 4.4:
Low-Emitting Materials,
Composite Wood and Agrifiber Products

PURPOSE

To limit the amount of _____, irritating and/or harmful contaminants used _____ to enhance the _____ and well-being of _____ and occupants.

REQUIREMENTS

_____ wood and agrifiber products must not contain any _____ formaldehyde resins.

Laminate _____ used to assemble on-site and shop-applied composite wood and agrifiber products must contain no added urea-_____ resins.

DOCUMENTATION

Contractor to submit a schedule of all composite _____ and agrifiber products used on the project and documentation proving no urea-formaldehyde _____ were used.

Composite wood and agrifiber products include _____, medium-density fiberboard (MDF), _____, wheatboard, strawboard, panel substrates, and _____ cores. _____ furniture and _____ is _____ to be included in this credit.

1 pt

RESPONSIBLE PARTY:
CONTRACTOR

EQ Credit 4.4:
Low-Emitting Materials,
Composite Wood and Agrifiber Products

ANSWER KEY

PURPOSE

To limit the amount of **odorous**, irritating and/or harmful contaminants used **indoors** to enhance the **comfort** and well-being of **installers** and occupants.

REQUIREMENTS

Composite wood and agrifiber products must not contain any **urea-**formaldehyde resins.

Laminate **adhesives** used to assemble on-site and shop-applied composite wood and agrifiber products must contain no added urea-**formaldehyde** resins.

DOCUMENTATION

Contractor to submit a schedule of all composite **wood** and agrifiber products used on the project and documentation proving no urea-formaldehyde **resins** were used.

Composite wood and agrifiber products include **particleboard**, medium-density fiberboard (MDF), **plywood**, wheatboard, strawboard, panel substrates, and **door** cores. **Systems** furniture and **seating** is **not** to be included in this credit.

CONSTRUCTION

EQ Credit 4.5:
Low-Emitting Materials,
Furniture and Seating

PURPOSE

To limit the amount of _____, irritating and/or harmful contaminants used _____ to enhance the _____ and well-being of _____ and occupants.

REQUIREMENTS

All _____ furniture and seating that was manufactured, _____, or refinished within _____ year prior to _____ must:

Option 1:
be _____ Indoor Air Quality certified.

Option 2:
not exceed the maximum indoor air concentrations for contaminants as determined by a procedure based on the EPA _____ Large Chamber Test Protocol for Measuring Emissions of VOCs and _____.

Option 3:
not exceed the maximum indoor air _____ for contaminants as determined by a procedure based on _____/_____ M7.1 - 2007 testing protocol.

DOCUMENTATION

Contractor to submit a schedule of all _____ furniture and seating (_____ and guest chairs) used on the project and documentation proving compliance.

REFERENCED STANDARDS

ANSI/BIFMA X7.1 - 2007, Standard for _____ and TVOC Emissions of Low-Emitting Office Furniture _____ and Seating

Environmental Technology _____ (ETV) Large Chamber Test Protocol for Measuring Emissions of VOCs and Aldehydes

_____ Certification Program

1 pt

RESPONSIBLE PARTY:
CONTRACTOR

EQ Credit 4.5: Low-Emitting Materials, Furniture and Seating

ANSWER KEY

PURPOSE

To limit the amount of **odorous**, irritating and/or harmful contaminants used **indoors** to enhance the **comfort** and well-being of **installers** and occupants.

REQUIREMENTS

All **systems** furniture and seating that was manufactured, **refurbished**, or refinished within **1** year prior to **occupancy** must:

Option 1:

    be **GREENGUARD** Indoor Air Quality certified.

Option 2:

    not exceed the maximum indoor air concentrations for contaminants as determined by a procedure based on the EPA **ETV** Large Chamber Test Protocol for Measuring Emissions of VOCs and **Aldehydes**.

Option 3:

    not exceed the maximum indoor air **concentrations** for contaminants as determined by a procedure based on **ANSI/BIFMA** M7.1 - 2007 testing protocol.

DOCUMENTATION

Contractor to submit a schedule of all **systems** furniture and seating (**task** and guest chairs) used on the project and documentation proving compliance.

REFERENCED STANDARDS

ANSI/BIFMA X7.1 - 2007, Standard for **Formaldehyde** and TVOC Emissions of Low-Emitting Office Furniture **Systems** and Seating

Environmental Technology **Verification** (ETV) Large Chamber Test Protocol for Measuring Emissions of VOCs and Aldehydes

GREENGUARD Certification Program

DESIGN

EQ Credit 5:
Indoor Chemical and Pollutant Source Control

PURPOSE

Reduce the _____ exposure to possible harmful _____ and chemical pollutants.

REQUIREMENTS

Install _____ entryway systems, such as _____, grates, and _____, at least _____ long in the direction of travel at building entrances.

Exhaust areas, such as high-volume copy areas, with _____ chemicals and gases. These areas should be located away from regularly occupied spaces.

Install MERV _____ or better filters at both _____ air and where outside air is supplied.

Provide for proper _____ and appropriate _____ of hazardous liquid wastes for housekeeping and laboratory spaces.

REFERENCED STANDARD

ASHRAE Standard _____ 1999: Method of Testing General Ventilation Air-_____ Devices for _____ Efficiency by _____ Size

1 pt

RESPONSIBLE PARTY:
INTERIOR DESIGNER

EQ Credit 5:
Indoor Chemical and Pollutant Source Control

ANSWER KEY

PURPOSE

Reduce the **occupant's** exposure to possible harmful **particulates** and chemical pollutants.

REQUIREMENTS

Install **permanent** entryway system, such as **grilles**, **grates**, and mats, at least **10'** long in the direction of travel at building entrances.

Exhaust areas, such as high-volume copy areas, with **hazardous** chemicals and gases. These areas should be located away from regularly occupied spaces

Install MERV **13** or better filters at both **return** air and where outside air is supplied.

Provide for proper **storage** and appropriate **disposal** of hazardous liquid wastes for housekeeping and laboratory spaces.

REFERENCED STANDARD

ASHRAE Standard **52.2** - 1999: Method of Testing General Ventilation Air-**Cleaning** Devices for **Removal** Efficiency by **Particle** Size

DESIGN

EQ Credit 6.1:
Controllability of Systems,
Lighting

PURPOSE

Increase the _____, comfort, and well-being for _____ by providing lighting system control to _____ occupants and groups in multioccupant spaces.

REQUIREMENTS

Provide individual lighting _____ for at least _____ percent of occupants and all _____ multioccupant spaces.

1 pt

RESPONSIBLE PARTY:
ELECTRICAL ENGINEER

EQ Credit 6.1:
Controllability of Systems,
Lighting

ANSWER KEY

PURPOSE

Increase the **productivity**, comfort, and well-being for **occupants** by providing lighting system control to **individual** occupants and groups in multioccupant spaces.

REQUIREMENTS

Provide individual lighting **controls** for at least **90** percent of occupants and all **shared** multioccupant spaces.

DESIGN

**EQ Credit 6.2:
Controllability of Systems,
Thermal Comfort**

PURPOSE

Increase the _____, comfort, and well-being for _____ by providing _____ comfort system control to _____ occupants and groups in multioccupant spaces.

REQUIREMENTS

Provide at least _____ percent of the building occupants and all of the shared _____-occupant spaces with thermal _____ controls.

Occupants must be able to control at least 1 of the _____ environmental factors of _____ comfort: air temperature, air _____, _____ temperature, and _____.

If _____ windows are used instead of controls, workstations must be located within _____ feet in front of and within _____ feet to either side. The window area must comply with ASHRAE _____ standards.

REFERENCED STANDARDS

_____ Standard _____ - 2007, _____ for Acceptable Indoor Air Quality

ASHRAE Standard _____ - 2004, _____ Environmental Conditions for Human _____

1 pt

RESPONSIBLE PARTY:
Mechanical Engineer

EQ Credit 6.2: Controllability of Systems, Thermal Comfort

ANSWER KEY

PURPOSE

Increase the **productivity**, comfort, and well-being for **occupants** by providing **thermal** comfort system control to **individual** occupants and groups in multioccupant spaces.

REQUIREMENTS

Provide at least **50** percent of the building occupants and all of the shared **multioccupant** spaces with thermal **comfort** controls.

Occupants must be able to control at least 1 of the **4** environmental factors of **thermal** comfort: air temperature, air **speed**, **radiant** temperature, and **humidity**.

If **operable** windows are used instead of controls, workstations must be located within **20** feet in front of and within **10** feet to either side. The window area must comply with ASHRAE **62.1** standards.

REFERENCED STANDARDS

ASHRAE Standard **62.1** - 2007, **Ventilation** for Acceptable Indoor Air Quality

ASHRAE Standard **55** - 2004, **Thermal** Environmental Conditions for Human **Occupancy**

DESIGN

**EQ Credit 7.1:
Thermal Comfort,
Design**

PURPOSE

Increase the _____ and _____-_____ for occupants by providing a comfortable _____ environment.

REQUIREMENTS

_____ systems must be designed to comply with ASHRAE _____ - 2004 and address the _____ occupancy factors (_____ rate and clothing _____) and the _____ building design factors (air _____, radiant temperature, air _____, and humidity).

REFERENCED STANDARD

ASHRAE Standard _____ - 2004, Thermal _____ Conditions for Human Occupancy

1 pt

RESPONSIBLE PARTY:
MECHANICAL ENGINEER

EQ Credit 7.1: Thermal Comfort, Design

ANSWER KEY

PURPOSE

Increase the **productivity** and **well-being** for occupants by providing a comfortable **thermal** environment.

REQUIREMENTS

HVAC systems must be designed to comply with ASHRAE **55** - 2004 and address the **2** occupancy factors (**metabolic** rate and clothing insulation) and the **four** building design factors (air **temperature**, radiant temperature, air **speed**, and humidity).

REFERENCED STANDARD

ASHRAE Standard **55** - 2004, Thermal **Environmental** Conditions for Human Occupancy

DESIGN

EQ Credit 7.2:
Thermal Comfort, Verification

PURPOSE

Assess the _____ comfort of the occupants over time.

REQUIREMENTS

Pursue and _____ EQ Credit _____: Thermal Comfort, _____.

Confirm commitment to conducting an _____ thermal comfort survey _____ to _____ months after occupancy and have a _____ action plan developed.

Install a permanent _____ system to _____ and _____ the comfort criteria.

REFERENCED STANDARD

_____ Standard 55 - 2004, _____ Environmental Conditions for Human _____

1 pt

RESPONSIBLE PARTY:

OWNER

EQ Credit 7.2:
Thermal Comfort, Verification

ANSWER KEY

PURPOSE

Assess the **thermal** comfort of the occupants over time.

REQUIREMENTS

Pursue and **earn** EQ Credit **7.1**: Thermal Comfort, **Design**.

Confirm commitment to conducting an **anonymous** thermal comfort survey **6** to **18** months after occupancy and have a **corrective** action plan developed.

Install a permanent **monitoring** system to **measure** and **record** the comfort criteria.

REFERENCED STANDARD

ASHRAE Standard 55 - 2004, **Thermal** Environmental Conditions for Human **Occupancy**

DESIGN

EQ Credit 8.1:
Daylight and Views,
Daylight

PURPOSE

For _____ occupied areas, introduce _____ and views to provide the building occupants with a _____ to the outdoor environment.

REQUIREMENTS

Provide daylight to _____ percent (1 point) or _____ percent (2 points) of the _____ occupied spaces.

COMPLIANCE PATH OPTIONS

OPTION 1: Simulation
Prove the spaces are daylit with between _____ and _____ footcandles.

OPTION 2: Prescriptive
Calculate the _____ zone for _____ lighting and/or _____ lighting strategies.

OPTION 3: Measurement
Record indoor light _____ levels to prove a minimum of _____ footcandles.

OPTION 4: Combination
Choose from Options 1 through _____ to prove compliance.

STRATEGIES

All strategies must include _____ control.

OPTION 2: Prescriptive

_____-lighting Daylight Zone:
Determine the _____ light _____ (VLT) for windows _____ above the floor. Multiply VLT by the _____ to-floor area ration (WFR) to prove a value between _____ and 0.180.

For ceiling obstructed daylight, the related floor area must be _____ from the calculations.

_____-lighting Daylight Zone:
Determine the daylight zone area under each _____ and add the _____ of the following three factors:

- _____ percent of the _____ height
- One-half the distance to the edge of the _____ skylight
- The distance to a fixed solid partition farther than _____ percent of the distance between the top of the wall and the ceiling

Skylight roof coverage must be between _____ and _____ percent of the roof area.

Skylights must have a minimum VLT of 0.5 and not be spaced apart more than _____ times the ceiling height.

If using a skylight _____ , it is required to have a measured haze value of at least _____ percent and a _____ _____ of sight to the diffuser must be avoided.

REFERENCED STANDARD

_____ D1003-07E1, Standard Test for Method for _____ and Luminous Transmittance of Transparent _____

1–2 pts

RESPONSIBLE PARTY:
ARCHITECT

EQ Credit 8.1: Daylight and Views, Daylight

ANSWER KEY

PURPOSE
For **regularly** occupied areas, introduce **daylight** and views to provide the building occupants with a **connection** to the outdoor environment.

REQUIREMENTS
Provide daylight to **75** percent (1 point) or **90** percent (2 points) of the **regularly** occupied spaces.

COMPLIANCE PATH OPTIONS
OPTION 1: **Simulation**
  Prove the spaces are daylit with between **25** and **500** footcandles.
OPTION 2: **Prescriptive**
  Calculate the **daylight** zone for **side**-lighting and/or **top**-lighting strategies.
OPTION 3: **Measurement**
  Record indoor light **footcandle** levels to prove a minimum of **25** footcandles.
OPTION 4: **Combination**
  Choose from Options 1 through **3** to prove compliance.

STRATEGIES
All strategies must include **glare** control.
OPTION 2: Prescriptive
  Side-lighting Daylight Zone:
    Determine the **visible** light **transmittance** (VLT) for windows **30"** above the floor. Multiply VLT by the **window**-to-floor area ration (WFR) to prove a value between **0.150** and 0.180.
    For ceiling obstructed daylight, the related floor area must be **excluded** from the calculations.
  Top-lighting Daylight Zone:
    Determine the daylight zone area under each **skylight** and add the lesser of the following three factors:
    - **70** percent of the ceiling height
    - One-half the distance to the edge of the **closest** skylight
    - The distance to a fixed solid partition farther than **70** percent of the distance between the top of the wall and the ceiling
    Skylight roof coverage must be between 3 and 6 percent of the roof area.
    Skylights must have a minimum VLT of 0.5 and not be spaced apart more than **1.4** times the ceiling height.
    If using a skylight **diffuser**, it is required to have a measured haze value of at least **90** percent and a **direct line** of sight to the diffuser must be avoided.

REFERENCED STANDARD
ASTM D1003-07E1, Standard Test for Method for **Haze** and Luminous Transmittance of Transparent **Plastics**

DESIGN

EQ Credit 8.2:
Daylight and Views,
Views for Seated Spaces

PURPOSE

For _____ occupied areas, introduce daylight and _____ to provide the building occupants with a _____ to the outdoor environment.

REQUIREMENTS

Provide direct _____ of sight to the _____ for at least _____ percent of regularly occupied areas.

Direct line of _____ is measured between _____ and _____ above the finish floor.

DOCUMENTATION

Provide a floor plan with _____ lines to _____ glazing.

Provide a section with direct line of sight to _____ glazing.

EXEMPLARY PERFORMANCE

Meet at least two out of the four measures:

- Provide at least _____ percent of regularly occupied areas with numerous lines of sight in multiple directions and at least _____ degrees apart

- Provide views for at least _____ percent of regularly occupied areas to at least 2 of the following: _____, human activity, or objects that are at least _____ feet from the glass

- Provide unobstructed views for at least _____ percent of regularly occupied areas for a distance of _____ times the head height of the vision glazing

- Provide views for at least _____ percent of regularly occupied areas that have a view factor of at least _____

1 pt

RESPONSIBLE PARTY:
INTERIOR DESIGNER

EQ Credit 8.2: Daylight and Views, Views for Seated Spaces

ANSWER KEY

PURPOSE

For **regularly** occupied areas, introduce daylight and **views** to provide the building occupants with a **connection** to the outdoor environment.

REQUIREMENTS

Provide direct **line** of sight to the **exterior** for at least **90** percent of regularly occupied areas.

Direct line of **sight** is measured between **30"** and **90"** above the finish floor.

DOCUMENTATION

Provide a floor plan with **sight** lines to **perimeter** glazing.

Provide a section with direct line of sight to **perimeter** glazing.

EXEMPLARY PERFORMANCE

Meet at least two out of the four measures:

- Provide at least **90** percent of regularly occupied areas with numerous lines of sight in multiple directions and at least **90** degrees apart

- Provide views for at least **90** percent of regularly occupied areas to at least 2 of the following: **vegetation**, human activity, or objects that are at least **70** feet from the glass

- Provide unobstructed views for at least 90 percent of regularly occupied areas for a distance of **3** times the head height of the vision glazing

- Provide views for at least **90** percent of regularly occupied areas that have a view factor of at least **3**

CHAPTER 9
INNOVATION IN DESIGN AND REGIONAL PRIORITY

THE PREVIOUS FIVE CHAPTERS DETAILED THE MAIN CATEGORIES of the Leadership in Energy and Environmental Design (LEED™) rating systems; this chapter focuses on the two bonus categories of the LEED for Commercial Interiors (CI) rating system: Innovation in Design (ID) and Regional Priority (RP). These categories are treated as bonus categories, as neither contains any prerequisites. The ID category encourages projects to explore new and innovative strategies and technologies, while the RP category offers additional point-earning opportunities focused on geographic environmental achievements.

Remember to switch back to the white flashcards for ID and RP topics.

INNOVATION IN DESIGN

The ID category encourages the exploration and implementation of new green building technologies, as well as exceeding the thresholds defined in the existing LEED credits. Remember, this is in addition to the SS Credit 1, Path 12 opportunity for one point. As shown in Table 9.1, the LEED CI rating system offers up to 6 points for projects within the ID category by addressing three different strategies:

1. Exemplary Performance
2. Innovation in Design
3. Including a LEED Accredited Professional on the project team

Create a flashcard to remember the three strategies to earn ID points.

ID Credit 1: Innovation or Exemplary Performance

Three of the available six ID points can be used toward the achievement of exemplary performance. As mentioned in the previous chapters, Exemplary Performance credits are achieved once projects surpass the minimum performance-based thresholds defined in the existing LEED credits, typically

Table 9.1 The Innovation in Design Category

D/C	Prereq/Credit	Title	Points
D/C	Credit 1	Innovation or Exemplary Performance	1 to 5
C	Credit 2	LEED Accredited Professional	1

243

the next incremental percentage threshold. For example, projects can earn Exemplary Performance credits (within the ID category) for achieving the following:

- Diverting 95 percent of construction waste per the requirements of MR Credit 2: Construction Waste Management
- Reducing water consumption within the tenant space by more than 45 percent per the requirements of WE Credit 1: Water Use Reduction
- When 30 percent or more of the total material cost includes products with recycled content per the requirements of MR Credit 4: Recycled Content

If the team uses all three opportunities for exemplary performance achievements, they still have at least one more point opportunity for implementing an innovative strategy. If the team pursues less than three credits within the ID category for exemplary performance, then they can pursue more innovative strategies. The teams should research credit interpretation requests (CIRs) to see if their proposed strategy has been incorporated or presented in the past, or issue a new CIR to inquire about the award potential. Successful ID solutions include strategies that demonstrate quantitative performance with environmental benefits and can be replicated by other project teams. Some examples previously submitted and awarded include:

- Incorporating cradle-to-cradle (C2C) certified products (Figure 9.1)
- Implementing an educational program for occupants and visitors (Figure 9.2)
- Using large amounts of fly ash in concrete
- Achieving LEED credits from other rating systems, such as a LEED CI project pursuing Acoustical Performance or Mold Prevention from the LEED for Schools™ rating system

 Remember the cradle-to-grave cycle from Chapter 7? C2C products are not only made of recycled products, but are recyclable after their useful life to avoid landfills and incineration facilities.

 Remember what fly ash is from Chapter 7?

Figure 9.1 C2C products help to extend the life of materials, although repurposed, reducing the need for virgin materials.
Photo courtesy of Steelcase, Inc.

Figure 9.2 Providing opportunities to educate the end-users and community about the benefits and strategies of green building helps to further transform the market and, therefore, contribute to earning LEED certification.
Photo courtesy of Rainwater HOG, LLC

When submitting for an Innovation in Design (ID) credit, each of the following four components need to be addressed for each credit being pursued:

1. Intent of strategy
2. Suggested requirements for compliance
3. Suggested documentation proving compliance with requirements
4. A narrative describing the strategy implemented

ID Credit 2: LEED Accredited Professional

The LEED rating systems also offer another point opportunity for including a LEED Accredited Professional (LEED AP) on the project team. Including a LEED AP on the project team can add efficiencies, as they are aware of the requirements of the LEED certification process. They are familiar with integrated design processes and understand how to evaluate the trade-offs and synergies of green building strategies and technologies. For the purposes of the exam, it is critical to remember that only **one point** can be awarded to projects for this credit; it does not matter how many LEED APs are on the project team, just as long as there is one. Note, LEED Green Associates do not qualify for the bonus point under this credit. The LEED AP credential certificate is required to prove compliance and earn the point.

REGIONAL PRIORITY

As seen in Table 9.2, the RP category offers the opportunity to earn up to four bonus points for achieving compliance of previously mentioned existing LEED credits. U.S. Green Building Council's (USGBC's) eight Regional Councils have consulted with the local chapters to determine which existing LEED credits are more challenging to achieve within certain zip codes. Based on the results of

 TIP Do you remember the point structure for LEED? How many points does a project need to achieve to earn Platinum status?

Table 9.2 The Regional Priority Category

D/C	Prereq/Credit	Title	Points
D/C	Credit 1	Regional Priority	1 to 4

their findings, USGBC compiled a database of all the zip codes in the United States (available on the USGBC website) and choose six existing LEED credits to coordinate with each corresponding geographic region. For example, a project located in Dania Beach, Florida, could earn a bonus point within the RP category for purchasing 20 percent of their materials that are extracted, processed, and manufactured within 500 miles from the project site, as USGBC has recognized that South Florida has a few opportunities to obtain building materials to comply with MR Credit 5: Regional Materials. A project could earn up to four Regional Priority credits (RPCs) out of the six opportunities presented. For the purposes of the exam, it is critical to remember that RPCs are not *new* credits.

QUIZ TIME!

Q9.1. Exemplary performance generally requires which of the following? (Choose one)

   A. Develop an innovative strategy not presented in any existing LEED credit.

   B. Achieve either 20 percent or the next incremental percentage threshold established by the existing LEED credit that is being exceeded, whichever is greater.

   C. Meet or exceed the next percentage threshold as listed within the existing credit.

   D. Surpass the defined threshold of an innovative strategy being proposed by the team.

   E. Achieve at least double the minimum effort described within the existing LEED credit, regardless of which credit is being exceeded.

Q9.2. Pursuing an Innovation in Design opportunity is appropriate when *at least one* of which of the following is true? (Choose two)

   A. The project is unable to meet the requirements established by an existing LEED credit.

   B. The compliance paths offered within an existing LEED credit are not possible to pursue.

   C. The project has exceeded or is projected to exceed the minimum performance established by an existing LEED credit.

   D. The project has achieved measurable performance in a LEED credit within another rating system.

Q9.3. Is it possible for the same building to earn multiple LEED certifications?

  A. Yes

  B. No

Q9.4. Which of the following statements are true regarding RPCs? (Choose three)

  A. Earning an RPC adds a bonus point to the project's total points.

  B. RPCs are new credits included in the LEED rating systems.

  C. Projects that are not registered with the 2009 versions of LEED are not awarded points within the Regional Priority category.

  D. Each zip code is assigned eight RPC opportunities.

  E. A project may earn up to four RPC bonus points.

Q9.5. How many points can be earned in the RP category? (Choose one)

  A. Six

  B. Three

  C. Two

  D. Four

  E. Ten

Q9.6. There are five LEED APs on the Botanical Center project, including three from the architectural firm, one from the mechanical engineering firm, and one from the electrical engineering firm. How many points can be achieved within the ID category for achieving this effort? (Choose one)

  A. One

  B. Two

  C. Three

  D. Four

Q9.7. What is the threshold to earn exemplary performance for EA Credit 4: Green Power? (Choose one)

  A. 65 percent

  B. 15 percent

  C. 75 percent

  D. 100 percent

  E. 20 percent

Q9.8. Which of the following is not in compliance with the exemplary performance requirements for EQ Credit 8.2: Daylight and Views, Views for Seated Spaces? (Choose three)

  A. At least 90 percent of regularly occupied spaces have multiple lines of sight to vision glazing at least 90 degrees apart.

B. At least 90 percent of regularly occupied areas have views to vegetation, human activity, and/or exterior objects at least 70 feet from the glazing.

C. At least 90 percent of regularly occupied spaces have unobstructed views.

D. At least 90 percent of the regularly occupied areas have a direct line of sight to the outdoor environment.

E. EQ Credit 8.2 is not eligible for exemplary performance.

F. Exemplary performance for this credit is awarded on a case-by-case basis.

G. At least 90 percent of regularly occupied spaces have access to views with a view factor of 3 or more.

Q9.9. Of the following MR credits, which one does not use cost to calculate compliance? (Choose one)

A. MR Credit 1.2: Building Reuse

B. MR Credit 3.1: Material Reuse

C. MR Credit 4: Recycled Content

D. MR Credit 5: Regional Materials

E. MR Credit 6: Rapidly Renewable Materials

Q9.10. Which of the following categories does not have a prerequisite? (Choose one)

A. Sustainable Sites

B. Water Efficiency

C. Materials and Resources

D. Energy and Atmosphere

E. Indoor Environmental Quality

Q9.11. Which of the following reference standards provides guidance for calculation space allotments for the storage and collection of recyclables? (Choose one)

A. ASHRAE 90.1

B. City of Seattle

C. ASTM 1907

D. Greenguard

E. ISO 14021

Q9.12. A project team is pursuing EA Credit 4: Green Power for a 44,000 square foot tenant improvement project. How much green power would they need to purchase in order to earn 5 points?

A. 44,000 kwh

B. 352,000 kwh

C. 704,000 kwh

D. 1,408,000 kwh

Q9.13. Which of the following strategies might earn a project team SS Credit 1: Site Selection, Path 5 Heat Island Effect, Roof? (Choose one)

    A. Roof material with an SRI value of 30 on a 3:12 roof

    B. Roof material with an SRI value of 17 on a 3:12 roof

    C. Roof material with an SRI value of 70 on a 2:12 roof

    D. Roof material with an SRI value of 29 on a 1:12 roof

Q9.14. Which of the following EQ credits requires a team to distinguish which spaces are classified as a regularly occupied space? (Choose four)

    A. EQ Credit 6.1: Controllability of Systems, Lighting

    B. EQ Credit 6.2: Controllability of Systems, Thermal Comfort

    C. EQ Credit 8.1: Daylighting and Views, Daylighting

    D. EQ Credit 2: Increased Ventilation

    E. EQ Credit 8.1: Daylighting and Views, Views for Seated Spaces

    F. EQ Credit 7.1: Thermal Comfort, Design

DESIGN

ID Credit 1:
Innovation in Design

PURPOSE

Allow for the opportunity to achieve _____ performance and/or _____ performance for green building strategies not addressed by the _____ Green Building _____ Systems.

STRATEGIES

PATH 1: Innovation in Design (Up to _____ points)

    1. Propose a _____ and innovative performance achieved with an environmental benefit.

    2. Proposed innovation must be applied comprehensively to the _____ project.

    3. Must be _____ to other projects and considerably better than sustainable design strategies.

PATH 2: Exemplary Performance (Up to 3 points)

    _____ the requirement of an existing LEED credit and/or achieve the stated _____ percentage _____.

DOCUMENTATION

PATH 1:

    _____ of proposed strategy

    _____ of proposed strategy

    _____ proving compliance with proposed requirements

    _____ to achieve innovative performance

1–5 pts

RESPONSIBLE PARTY:
PROJECT TEAM

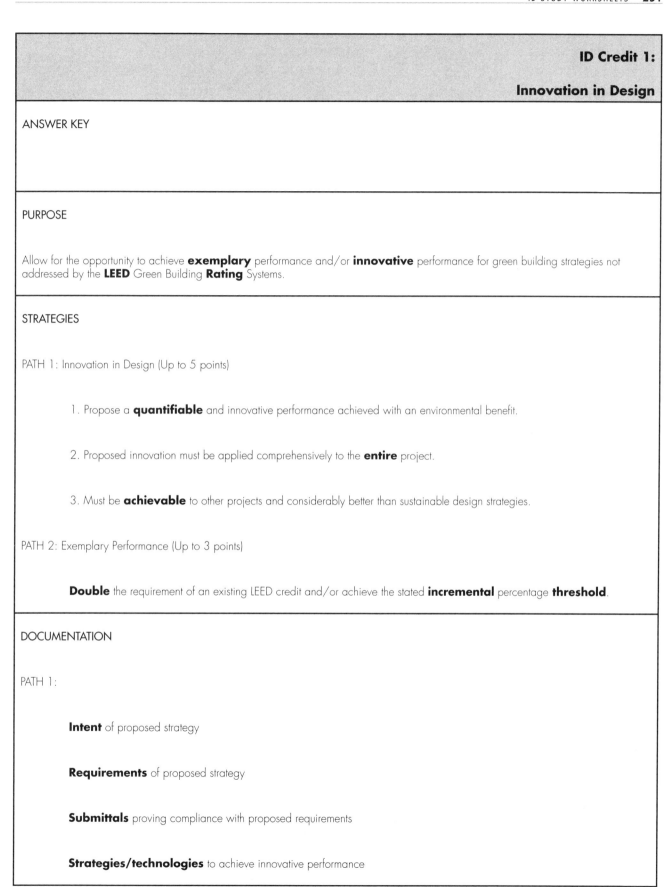

CONSTRUCTION

ID Credit 2:
LEED Accredited Professional (AP)

PURPOSE

_____ the _____ and _____ process by encouraging the integration required by the LEED Green Building Rating Systems.

REQUIREMENTS

Engage at least _____ LEED AP as a integral project team member.

REFERENCED STANDARDS

LEED _____ Professional (AP)

Green Building _____ Institute (GBCI), www.gbci.org

1 pt

RESPONSIBLE PARTY:

PROJECT TEAM

ID STUDY WORKSHEETS

ID Credit 2: LEED Accredited Professional (AP)

ANSWER KEY

PURPOSE

Streamline the **application** and **certification** process by encouraging the integration required by the LEED Green Building Rating Systems.

REQUIREMENTS

Engage at least **one** LEED AP as a integral project team member.

REFERENCED STANDARDS

LEED **Accredited** Professional (AP)

Green Building **Certification** Institute (GBCI), www.gbci.org

PART III
STUDY TIPS AND APPENDICES

CHAPTER 10
STUDY TIPS

AS MENTIONED EARLIER, IN THE INTRODUCTION OF THIS BOOK, this chapter is dedicated to providing an approach for the rest of your study efforts. It includes tips for taking online practice exams and resources on where to find additional information while you continue to study, as well as providing an insight to the Prometric testing center environment and the exam format structure.

PREPARING FOR THE LEED AP ID+C EXAM: WEEK EIGHT

By the time you read this section, it should be Week Eight of your study efforts. You should have your white set of flashcards covering the basics of Leadership in Energy and Environmental Design (LEED®) (including Innovation in Design [ID] and Regional Priority [RP] bonus categories) and your color-coded cards separated into the five main categories of the LEED rating systems. This week will be a great opportunity to rewrite your cheat sheet at least three times. Note that your cheat sheet may evolve as you take a few online practice exams.

During Week Eight, you may need to refer to additional resources while studying. For example, if you want to learn more about the cost implications of LEED projects, refer to *Cost of Green Revisited* by Davis Langdon. Although a sample credit is provided in Appendix M, I would recommend downloading the LEED for Commercial Interiors rating system from the U.S. Green Building Council (USGBC) website and skimming through it to see how the categories, prerequisites, and credits are organized and presented. I would also recommend reading through *Guidance on Innovation & Design (ID) Credits* and downloading and reviewing the sample credits from the USGBC website. All of these references are available to download from the *LEED AP Interior Design + Construction Candidate Handbook* on the GBCI website for free. Again, it is highly recommended that you download the most current candidate handbook from the GBCI website, but as a point of reference, at the time of printing the references included:

- *LEED for Interior Design and Construction Reference Guide* (USGBC)
- *Cost of Green Revisited,* by Davis Langdon (2007)
- *Sustainable Building Technical Manual: Part II* by Anthony Bernheim and William Reed (1996)
- *Guidance on Innovation & Design (ID) Credits* (USGBC, 2004)
- *Guidelines for CIR Customers* (USGBC, 2007)
- LEED-Online—sample credit templates (www.usgbc.org)

TIP Download and read through two of the references from the *LEED AP Interior Design + Construction Candidate Handbook*:
- Sample Credits
- Guidance on Innovation in Design (ID) Credits

Some other resources include:

- *www.usgbc.org.* You may want to check out some of the rating system scorecards or read about credit weightings of LEED v3. The USGBC website is also your primary source to learn about any updates to the LEED rating systems.
- *www.gbci.org.* Make sure you download the current candidate handbook, and you may want to reference the disciplinary policy and the minimum program requirements and the project registration information.
- *www.leedonline.com.* Even if you do not have any projects assigned to you, you will still be able to see what it looks like and watch a demo video.

Practice Exam Approach

Also during Week Eight, you should take some online practice exams. Although there are many sample exam questions provided in this book, it is helpful to practice for the real-life testing environment scenario. Search online, as you will find that there are a few options from which to choose for online practice exams, including www.GreenEDU.com. When you are taking a practice exam, pretend it is the real thing. For example, time yourself, have scratch paper and a pencil available, make a cheat sheet in about two to three minutes, do not use this book or your flashcards, and avoid any disruptions. Some of the online practice exams allow you to flag questions you are doubtful about to remind you to go back and answer them later. Take advantage of this for practice, as the real test is formatted in the same manner.

Most of the questions include multiple choices with multiple answers required. When approaching these types of questions, you are best advised to "hide" or "cover up" the provided answer options with your hand or a sheet of paper and formulate your own answers to help avoid getting sidetracked by the answer selection choices. Once you uncover or reveal the answer choices, make sure to read through all of the options before selecting your final answer(s). Be sure to read each question carefully and select the proper number of answers. After taking the practice exam, go through the answer key and evaluate your score. On the first practice exam, read through each question and answer one by one to understand how you decided on the correct answer and where you went wrong on the incorrect ones. Try to notice a pattern on your strengths and weaknesses to determine where your study efforts need to be devoted to improve your score. After taking the first practice exam, you may just want to focus on the questions you answered incorrectly.

THE TESTING CENTER ENVIRONMENT

The introduction of this book described the opportunity to make a cheat sheet after you completed the tutorial at the testing center, and Chapter 1 detailed how to schedule your exam date with a Prometric testing center. Hopefully, your exam date is still not scheduled at this point, as one more week of preparation time is suggested to review your flashcards, to refine your cheat sheet, and to give you the opportunity to take a few online practice exams. As stated earlier, it is best to assess your knowledge before scheduling your exam date.

During Week Nine, the week of your test date, there are a few things to remember before you sit for the exam:

- Remember to visit the GBCI website and download the latest version of the *LEED AP Interior Design + Construction Candidate Handbook*.
- Confirm your exam date at least one day prior.
- Find the Prometric testing center and map your path to make sure you know where you are going on your exam day.
- Keep rewriting your cheat sheet and studying your flashcards. Take your flashcards everywhere with you!

To be prepared on the day of your exam, please note the following:

- Bring your picture ID with matching name, just as it is on your GBCI profile.
- Dress comfortably and bring a sweater or a jacket, as the testing center may be cold.
- Be sure to get plenty of rest and eat something, as you will not want to take any breaks during the exam to grab a bite or a drink (the clock cannot be paused).
- Be sure to **check in** at least 30 minutes prior to your testing time. If you miss your scheduled exam time, you will be considered absent and will have to forfeit your exam fees and the opportunity to take the exam.
- Be sure to use the restroom after checking in and prior to being escorted to your workstation. Remember, no breaks!
- You will be observed during your testing session and will be audio and video recorded as well.
- You will not be allowed to bring any personal items to your workstation, such as a calculator, paper, pencils, a purse, a wallet, food, or books.

EXAM STRUCTURE

The exam is structured to test you on three components, as described in the candidate handbook provided by GBCI. You will be tested on recognition items, application items, and analysis items. The recognition items test your ability to remember factual data once presented in a similar environment to the exam references. For example, you may need to provide the definition for a term or recall a fact. The application items present a situation for you to solve using the principles and elements described in the exam format. These questions may require you to perform a calculation or provide the process or sequence of actions (i.e., CIRs, registration, certification). The analysis items are presented to determine your ability to evaluate a problem and create a solution. These question types are more challenging, as you must be able to decipher the different components of the problem and also assess the relationships of the components.

 TIP A calculator will be provided should you be required to perform any calculations.

The exam questions are separated into categories of focus areas and then coordinated with an applicable rating system category. For example, project site factors coordinate with the Sustainable Sites (SS) category, and water management issues coordinate with the Water Efficiency (WE) category. Project systems and energy impacts coordinate with the Energy & Atmosphere (EA) category, whereas

 TIP Remember, the exam is composed of multiple-choice questions. No written answers are required!

acquisition, installation, and management of project materials coordinate with the Materials & Resources (MR) category. Improvements to the indoor environment coordinate with the Indoor Environmental Quality (EQ) category. Stakeholder involvement in innovation, project surroundings, and public outreach coordinate with the Innovation in Design (ID) and Regional Priority (RP) categories. Therefore, you should be familiar with each of the credit categories as presented earlier in Part II, Chapters 4–9.

When at the Testing Center

To give you an idea of what to expect, once you are at your workstation:

- You should dedicate 2 hours and 20 minutes to take the exam:
    - 10-minute tutorial
    - 2-hour exam
    - 10-minute exit survey
- The tutorial is computer based, so make sure your workstation's monitor, keyboard, and mouse are all functioning properly. After completing the tutorial, remember to create your cheat sheet in the time left over.
- The two-hour exam is composed of 100 multiple-choice questions. Just as with the practice exam questions, in order for the question to be counted as CORRECT, you must select **all** of the correct answers within each question, as there is no partial credit for choosing two out of the three correct answers.
- Although some of the practice exam questions in this book are formatted with a true or false statement or "All of the above" as an answer selection, you are less likely to find this on the real exam, as the questions tend to be straightforward and clear, to avoid any confusion.
- You will not see any credit numbers listed on their own, as all credit names will include the full name.
- Appendix J includes a list of commonly used acronyms. Although most of them are spelled out on the exam, it is still helpful to know what they are!
- During the exam, you will have the opportunity to mark or flag questions to come back to later. It is advised that you take advantage of this, as you may be short on time and want to revisit only the questions you were doubtful about. Note that any unanswered questions are marked INCORRECT, so it is best to at least try.
- The 10-minute exit survey is followed by your exam results—yes, instant and immediate results!

> **TIP** Remember to rely on your instincts. Typically, the first answer that comes to mind is often the right one!

Exam Scoring

The exams are scored on a scale from 125 to 200, where 170+ is considered passing. Do not worry about how the questions are weighted, just do your best! Should you need to retake the exam, your application is valid for one year, and therefore you will have three chances within the year to earn a score of 170 or more. Consult the candidate handbook for more information.

After the Exam

Once you have passed the LEED AP ID+C exam, remember to change your signature to reflect earning the credential! Although your certificate will not arrive immediately, remember, you must fulfill 30 hours of continuing education units over the next two years, including six LEED-specific continuing education units. The two-year reporting period begins the same day of your exam. Refer to the *Credential Maintenance Program (CMP)* handbook found on the GBCI website, for more information. There is also a code of conduct you must abide by, as stipulated in the disciplinary policy posted on the GBCI website at www.gbci.org/Files/Disc_ExamAppeals_Policy.pdf. It states that individuals with LEED credentials must:

A. Be truthful, forthcoming, and cooperative in their dealings with GBCI.

B. Be in continuous compliance with GBCI rules (as amended from time to time by GBCI).

C. Respect GBCI intellectual property rights.

D. Abide by laws related to the profession and to general public health and safety.

E. Carry out their professional work in a competent and objective manner.

TIP The disciplinary policy found on the GBCI website also includes the exam appeals policy, in case it is needed.

Appendix A

LEED CI RATING SYSTEM

Overview of the LEED CI Rating System

REFERENCE GUIDE	RATING SYSTEM	APPLICABLE PROJECT TYPES
ID+C	**LEED for Commercial Interiors (CI)**	
		• For tenant spaces:
		– office, retail & institutional
		– tenant spaces that don't occupy entire building
		• Works hand-in-hand with LEED CS
	LEED for Retail: Commercial Interiors	

Appendix B

MINIMUM PROGRAM REQUIREMENTS (MPRs) FOR LEED® CI

Minimum Program Requirements

1	**MUST COMPLY WITH ENVIRONMENTAL LAWS**
	Must comply with all applicable federal, state, and local building-related environmental laws and regulations in place where the project is located.
	This condition must be satisfied from the date of or the initiation of schematic design, whichever comes first, until the date that the building receives a certificate of occupancy or similar official indication that it is ready for use.
2	**MUST BE A COMPLETE, PERMANENT BUILDING OR SPACE**
	All LEED projects must be designed for, constructed on, and operated on a permanent location on already existing land.
	No building or space that is designed to move at any point in its lifetime may pursue LEED certification.
	The LEED project scope must include a complete interior space distinct from other spaces within the same building with regards to ownership, management, lease, or party wall separation.
	Construction prerequisites and credits may not be submitted for review until substantial completion of construction has occurred.
3	**MUST USE A REASONABLE SITE BOUNDARY**
	1. The LEED project boundary must include all contiguous land that is associated with and supports normal building operations for the LEED project building, including all land that was or will be disturbed for the purpose of undertaking the LEED project.
	2. The LEED project boundary may not include land that is owned by a party other than that which owns the LEED project unless that land is associated with and supports normal building operations for the LEED project building.
	3. LEED projects located on a campus must have project boundaries such that if all the buildings on campus become LEED certified, then 100% of the gross land area on the campus would be included within a LEED boundary. If this requirement is in conflict with MPR 7, Must Comply with Minimum Building Area to Site Area Ratio, then MPR 7 will take precedence.
	4. Any given parcel of real property may be attributed to only a single LEED project building.
	5. Gerrymandering of a LEED project boundary is prohibited: the boundary may not unreasonably exclude sections of land to create boundaries in unreasonable shapes for the sole purpose of complying with prerequisites or credits.
	6. If any land was or will be disturbed for the purpose of undertaking the LEED project, then that land must be included within the LEED project boundary.

4	**MUST COMPLY WITH MINIMUM FLOOR AREA REQUIREMENTS**
	The LEED project must include a minimum of 250 square feet of gross floor area.
5	**MUST COMPLY WITH MINIMUM OCCUPANCY RATES**
	Full-time equivalent occupancy
	The LEED project must serve one or more full-time equivalent (FTE) occupant(s), calculated as an annual average in order to use LEED in its entirety. If the project serves less than one annualized FTE, optional credits from the Indoor Environmental Quality category may not be earned (the prerequisites must still be earned).
6	**COMMITMENT TO SHARE WHOLE-BUILDING ENERGY AND WATER USAGE DATA**
	All certified projects must commit to sharing with USGBC and/or GBCI all available actual whole-project energy and water usage data for a period of at least five years. This period starts on the date that the LEED project begins typical physical occupancy if certifying under NC, CS, Schools, or CI, or the date that the building is awarded certification if certifying under EBOM.
	This commitment must carry forward if the building or space changes ownership or lessee.
7	**MUST COMPLY WITH A MINIMUM BUILDING AREA-TO-SITE AREA RATIO**
	The gross floor area of the LEED project building must be no less than 2% of the gross land area within the LEED project boundary.

Appendix C
LEED® CERTIFICATION PROCESS

The Basic Steps in the LEED Certification Process

PROJECT REGISTRATION	
	Access to LEED Online
	– LEED Scorecard
	– LEED Credit Submittal Templates
DESIGN APPLICATION PHASE (OPTIONAL)	
	Submit Credits and Prerequisites via LEED-Online
	– Comes back "Anticipated" or "Denied" (25 days)
	– No points awarded
	Clarification Request (25 days)
	Final Design Review (15 days)
	– Project team can:
	– ACCEPT: goes to Construction Application phase
	– APPEAL: goes to Design Appeal phase
DESIGN APPEAL PHASE	
	Changes made and submitted once again
	– Comes back "Anticipated" or "Denied" (25 days)
	No clarification requests
	Final Design Review (15 days)
CONSTRUCTION APPLICATION PHASE	
	Submit via LEED-Online (both design and construction)
	– Comes back "Anticipated" or "Denied" (25 days)
	– No points awarded yet
	Clarification Request (25 days)
	Final Construction Review (15 days)
	– Project team can:
	– ACCEPT: goes to Certification/Denial phase
	– APPEAL: goes to Construction Appeal phase

CONSTRUCTION APPEAL PHASE	
	Changes made and submitted once again
	– Comes back "Anticipated" or "Denied"
	No clarification requests
	Final Construction Review (15 days)

CERTIFIED/DENIAL PHASE
After Final Construction Review is ACCEPTED:
– Certification Level Awards: Certified, Silver, Gold, or Platinum
– Denied: project closed (appeals should be done in prior phases)

Appendix D
MAIN CATEGORY SUMMARIES

Sustainable Sites
  Site Selection
- SS Credit 1: Site Selection
  - Path 1: Brownfield Redevelopment
  - Path 2: Stormwater Design, Quantity Control
  - Path 3: Stormwater Design, Quality Control
  - Path 4: Heat Island Effect, Nonroof
  - Path 5: Heat Island Effect, Roof
  - Path 6: Light Pollution Reduction
  - Path 7: Water Efficient Landscaping—50 percent Reduction
  - Path 8: Water Efficient Landscaping—100 percent Reduction
  - Path 9: Innovative Wastewater Technologies
  - Path 10: Water Use Reduction—30 percent
  - Path 11: On-Site Renewable Energy
  - Path 12: Other Quantifiable Environmental Performance
- SS Credit 2: Development Density and Community Connectivity

  Transportation
- SS Credit 3.1: Public Transportation Access
- SS Credit 3.2: Bicycle Storage and Changing Rooms
- SS Credit 3.3: Parking Capacity

Water Efficiency
  Indoor Water Use Reduction
- WE Prerequisite 1: Water Use Reduction—20 percent
- WE Credit 1: Water Use Reduction

Energy & Atmosphere
  Energy Efficiency
- EA Prerequisite 2: Minimum Energy Performance
- EA Credit 1.1: Optimize Energy Performance—Lighting Power
- EA Credit 1.2: Optimize Energy Performance—Lighting Controls
- EA Credit 1.3: Optimize Energy Performance—HVAC
- EA Credit 1.4: Optimize Energy Performance—Appliances and Equipment

  Tracking Energy Consumption
- EA Prerequisite 1: Fundamental Commissioning of Building Energy Systems
- EA Credit 2: Enhanced Commissioning
- EA Credit 3: Measurement and Verification

  Managing Refrigerants
- EA Prerequisite 3: Fundamental Refrigerant Management

Renewable Energy
- EA Credit 4: Green Power

Materials & Resources
Salvaged Materials and Material Reuse
- MR Credit 1.2: Building Reuse, Maintain Interior Nonstructural Elements
- MR Credit 3.1: Materials Reuse
- MR Credit 3.2: Materials Reuse, Furniture and Furnishings

Building Material Selection
- MR Credit 4: Recycled Content
- MR Credit 5: Regional Materials
- MR Credit 6: Rapidly Renewable Materials
- MR Credit 7: Certified Wood

Waste Management
- MR Prerequisite 1: Storage and Collection of Recyclables
- MR Credit 1.1: Tenant Space, Long-Term Commitment
- MR Credit 2: Construction Waste Management

Indoor Environmental Quality
Indoor Air Quality
Ventilation
- EQ Prerequisite 1: Minimum Indoor Air Quality Performance
- EQ Credit 1: Outdoor Air Delivery Monitoring
- EQ Credit 2: Increased Ventilation

IAQ Practices during Construction
- EQ Credit 3.1: Construction IAQ Management, During Construction
- EQ Credit 3.2: Construction IAQ Management, Before Occupancy
- EQ Credit 4.1: Low-Emitting Materials, Adhesives and Sealants
- EQ Credit 4.2: Low-Emitting Materials, Paints and Coatings
- EQ Credit 4.3: Low-Emitting Materials, Flooring Systems
- EQ Credit 4.4: Low-Emitting Materials, Composite Wood and Agrifiber Products
- EQ Credit 4.5: Low-Emitting Materials, Systems Furniture and Seating

Prevention and Segregation Methods
- EQ Prerequisite 2: Environmental Tobacco Smoke (ETS) Control
- EQ Credit 5: Indoor Chemical and Pollutant Source Control

Thermal Comfort
- EQ Credit 6.2: Controllability of Systems, Thermal Comfort
- EQ Credit 7.1: Thermal Comfort, Design
- EQ Credit 7.2: Thermal Comfort, Verification

Lighting
- EQ Credit 6.1: Controllability of Systems, Lighting
- EQ Credit 8.1: Daylight and Views, Daylight
- EQ Credit 8.2: Daylight and Views, Views for Seated Spaces

Appendix E

RELATED PREREQUISITES AND CREDITS

Category	Item	Description	Credit 1	SS C1 Path 1	SS C1 Path 2	SS C1 Path 3	SS C1 Path 4	SS C1 Path 5	SS C1 Path 6	SS C1 Path 7	SS C1 Path 8	SS C1 Path 9	SS C1 Path 10	SS C1 Path 11	SS C1 Path 12	Credit 2	Credit 3.1	Credit 3.2	Credit 3.3
Sustainable Sites	Credit 1	Site Selection																	
	Credit 1 Path 1	Brownfield Redevelopment																	
	Credit 1 Path 2	Stormwater Design, Quantity Control				X	X	X		X	X		X						
	Credit 1 Path 3	Stormwater Design, Quality Control			X		X	X		X	X		X						
	Credit 1 Path 4	Heat Island Effect, Non-Roof			X	X				X	X								
	Credit 1 Path 5	Heat Island Effect, Roof			X	X				X	X		X						
	Credit 1 Path 6	Light Pollution Reduction																	
	Credit 1 Path 7	Water Efficient Landscaping - Reduce by 50%			X	X	X	X			X								
	Credit 1 Path 8	Water Efficient Landscaping - Reduce by 100%			X	X	X	X					X	X	X	X			
	Credit 1 Path 9	Innovative Wastewater Technologies											X						
	Credit 1 Path 10	Water Use Reduction, 30% Reduction			X	X		X		X	X	X							
	Credit 1 Path 11	On-Site Renewable Energy											X	X		X			
	Credit 1 Path 12	Other Quantifiable Environmental Performance																	
	Credit 2	Development Density & Community Connectivity																X	
	Credit 3.1	Alternative Transportation, Public Transportation Access																	X
	Credit 3.2	Alternative Transportation, Bicycle Storage & Changing Rooms																		X
	Credit 3.3	Alternative Transportation, Parking Capacity																		
WE	Prerequisite 1	Water Use Reduction, 20% Reduction			X	X				X	X	X	X						
	Credit 1	Water Use Reduction			X	X				X	X	X	X						
Energy & Atmosphere	Prerequisite 1	Fundamental Commissioning of the Building Energy Systems																	
	Prerequisite 2	Minimum Energy Performance																	
	Prerequisite 3	Fundamental Refrigerant Management																	
	Credit 1.1	Optimize Energy Performance, Lighting Power																	
	Credit 1.2	Optimize Energy Performance, Lighting Controls																	
	Credit 1.3	Optimize Energy Performance, HVAC																	
	Credit 1.4	Optimize Energy Performance, Equipment & Appliances																	
	Credit 2	Enhanced Commissioning																	
	Credit 3	Measurement & Verification																	
	Credit 4	Green Power																	
Materials & Resources	Prerequisite 1	Storage & Collection of Recyclables																	
	Credit 1.1	Tenant Space - Long-Term Commitment	X															X	
	Credit 1.2	Building Reuse, Maintain 50% of Interior Non-Structural Elements																	
	Credit 2	Construction Waste Management		X															
	Credit 3.1	Materials Reuse																	
	Credit 3.2	Materials Reuse, Furniture & Furnishings																	
	Credit 4	Recycled Content																	
	Credit 5	Regional Materials																	
	Credit 6	Rapidly Renewable Materials																	
	Credit 7	Certified Wood																	
Indoor Environmental Quality	Prerequisite 1	Minimum IAQ Performance		X														X	X	X
	Prerequisite 2	Environmental Tobacco Smoke (ETS) Control																	
	Credit 1	Outdoor Air Delivery Monitoring		X														X	X	X
	Credit 2	Increased Ventilation																	
	Credit 3.1	Construction IAQ Mgmt Plan: During Construction																	
	Credit 3.2	Construction IAQ Mgmt Plan: Before Occupancy																	
	Credit 4*	Low-Emitting Materials																	
	Credit 5	Indoor Chemical & Pollutant Source Control																	
	Credit 6.1	Controllability of Systems, Lighting																	
	Credit 6.2	Controllability of Systems, Thermal Comfort																	
	Credit 7.1	Thermal Comfort, Design																	
	Credit 7.2	Thermal Comfort, Verification																	
	Credit 8.1	Daylighting & Views: Daylight 75% of Spaces																	
	Credit 8.2	Daylighting & Views: Views for 90% of Spaces																	

*: Low-emitting material credits are all interrelated

APPENDIX E **271**

Appendix F
SAMPLE LEED CI SCORECARD[1]

Project Name
Date

LEED 2009 for New Commercial Interiors
Project Checklist

Yes	?	No	Sustainable Sites		Possible Points:	21
			Credit 1	Site Selection		1 to 5
			Credit 2	Development Density and Community Connectivity		6
			Credit 3.1	Alternative Transportation—Public Transportation Access		6
			Credit 3.2	Alternative Transportation—Bicycle Storage and Changing Rooms		2
			Credit 3.3	Alternative Transportation—Parking Availability		2

			Water Efficiency		Possible Points:	11
Y			Prereq 1	Water Use Reduction – 20% Reduction		
			Credit 1	Water Use Reduction		6 to 11

			Energy & Atmosphere		Possible Points:	37
Y			Prereq 1	Fundamental Commissioning of Building Energy Systems		
Y			Prereq 2	Minimum Energy Performance		
Y			Prereq 3	Fundamental Refrigerant Management		
			Credit 1.1	Optimize Energy Performance—Lighting Power		1 to 5
			Credit 1.2	Optimize Energy Performance—Lighting Controls		1 to 3
			Credit 1.3	Optimize Energy Performance—HVAC		5 to 10
			Credit 1.4	Optimize Energy Performance—Equipment and Appliances		1 to 4
			Credit 2	Enhanced Commissioning		5
			Credit 3	Measurement & Verification		2 to 5
			Credit 6	Green Power		5

			Materials & Resources		Possible Points:	14
Y			Prereq 1	Storage & Collection of Recyclables		
			Credit 1.1	Tenant Space—Long-Term Commitment		1
			Credit 1.2	Building Reuse		1 to 2
			Credit 2	Construction Waste Management		1 to 2
			Credit 3.1	Materials Reuse		1 to 2
			Credit 3.2	Materials Reuse—Furniture and Furnishings		1
			Credit 4	Recycled Content		1 to 2
			Credit 5	Regional Materials		1 to 2
			Credit 6	Rapidly Renewable Materials		1
			Credit 7	Certified Wood		1

			Indoor Environmental Quality		Possible Points:	17
Y			Prereq 1	Minimum IAQ Performance		
Y			Prereq 2	Environmental Tobacco Smoke (ETS) Control		
			Credit 1	Outdoor Air Delivery Monitoring		1
			Credit 2	Increased Ventilation		1
			Credit 3.1	Construction IAQ Management Plan – During Construction		1
			Credit 3.2	Construction IAQ Management Plan – Before Occupancy		1
			Credit 4.1	Low-Emitting Materials, Adhesives & Sealants		1
			Credit 4.2	Low-Emitting Materials, Paints & Coatings		1
			Credit 4.3	Low-Emitting Materials, Flooring Systems		1
			Credit 4.4	Low-Emitting Materials—Composite Wood and Agrifiber Products		1
			Credit 4.5	Low-Emitting Materials—Systems Furniture and Seating		1
			Credit 5	Indoor Chemical & Pollutant Source Control		1
			Credit 6.1	Controllability of Systems—Lighting		1
			Credit 6.2	Controllability of Systems—Thermal Comfort		1
			Credit 7.1	Thermal Comfort—Design		1
			Credit 7.2	Thermal Comfort—Verification		1
			Credit 8.1	Daylight and Views—Daylight		1 to 2
			Credit 8.2	Daylight & Views, Views		1

			Innovation and Design Process		Possible Points:	6
			Credit 1.1	Innovation in Design: Specific Title		1
			Credit 1.2	Innovation in Design: Specific Title		1
			Credit 1.3	Innovation in Design: Specific Title		1
			Credit 1.4	Innovation in Design: Specific Title		1
			Credit 1.5	Innovation in Design: Specific Title		1
			Credit 2	LEED Accredited Professional		1

			Regional Priority Credits		Possible Points:	4
			Credit 1.1	Regional Priority: Specific Credit		1
			Credit 1.2	Regional Priority: Specific Credit		1
			Credit 1.3	Regional Priority: Specific Credit		1
			Credit 1.4	Regional Priority: Specific Credit		1

			Total		Possible Points:	110

Certified: 40 to 49 points, **Silver:** 50 to 59 points, **Gold:** 60 to 79 points, **Platinum:** 80 to 110 points

Appendix G

EXEMPLARY PERFORMANCE OPPORTUNITIES – LEED® CI V3.0

Sustainable Sites		
SS Credit 1 Option 2: Path 4*	Heat Island Effect: Non-Roof (comply with 2 or more strategies)	Provide Shade, light paving, or open grid for 30% of hardscape
		Place a minimum of 50% of parking spaces under cover
		Use an open-grid pavement system (<50% impervious) for 50% of the parking lot area
SS Credit 1 Option 2: Path 5*	Heat Island Effect: Roof	100% of project area consists of a vegetated roof system
SS Credit 1 Option 2: Path 10*	Water Use Reduction	Reduce whole building water consumption by at least 40%
SS Credit 1 Option 2: Path 11*	Onsite Renewable Energy	Use on-site renewable energy systems supplying at least 10% of the annual building energy cost
SS Credit 2	Development Density & Community Connection	1. The project building itself must have a density at least double that of the average density within the calculated area, OR 2. The average density within an area twice as large as that for the base credit achievement must be at least 120,000 square feet per acre.
SS Credit 3.1	Alternative Transportation, Public Transportation Access	1. Institute a comprehensive transportation management plan that demonstrates a quantifiable reduction in automobile use
		2. Double Transit Ridership: Locate the tenant space in building that is within 1/2 mile of at least 2 existing rail, light rail, or subway stations, OR locate the tenant space in a building that is within 1/4 mile of at least 2 or more stops for 4 or more public or campus (private) bus lines usable by tenants, AND Frequency of service must be least 200 transit rides per day, total, at these stops
SS Credit 3.2	Alternative Transportation, Bicycle Storage and Changing Rooms	Institute a comprehensive transportation management plan that demonstrates a quantifiable reduction in automobile use

| SS Credit 3.3 | Alternative Transportation, Parking Availability | Institute a comprehensive transportation management plan that demonstrates a quantifiable reduction in automobile use |

Water Efficiency		
WE Credit 1	Water Use Reduction	45% in projected potable water use for tenant space

Energy & Atmosphere		
EA Credit 1.1	Optimize Energy Performance – Lighting Power	Reduce lighting power density to 40% or more below the standard
EA Credit 1.2	Optimize Energy Performance – Lighting Controls	Implementing daylight responsive controls for 75% of the connected lighting load or by installing occupancy-responsive controls for 95% of the connected lighting load
EA Credit 1.3	Optimize Energy Performance – HVAC (Option 2)	Demonstrate HVAC component performance for the tenant space is 33% more efficient than a system that is in minimum compliance with ASHRAE 90.1
EA Credit 1.4	Optimize Energy Performance – Equipment and Appliances	Achieve a rated power of 97% attributable to ENERGY STAR – qualified equipment and appliances
EA Credit 4	Green Power	Purchase 100% of the building's calculated annual energy use from contacted green power (or a default of 16kWh per square foot per year)

Materials & Resources		
MR Credit 1.2	Building Reuse, Maintain Interior Nonstructural Components	Reuse at least 80% (by area) of existing non-structural elements
MR Credit 2	Construction Waste Management	Divert ≥ 95% of total waste from disposal
MR Credit 3.1	Material Reuse	15% or more of total materials cost
MR Credit 3.2	Material Reuse, Furniture and Furnishings	60% of more of total furniture and furnishings budget
MR Credit 4	Recycled Content	30% or more of total materials cost
MR Credit 5	Regional Materials	20% of total materials cost, regionally harvested, extracted, and manufactured
MR Credit 6	Rapidly Renewable Materials	10% or more of total materials cost
MR Credit 7	Certified Wood	95% or more of total new wood building components cost

Indoor Environmental Quality		
EQ Credit 8.2	Daylight and Views, Views for Seated Spaces	1. 90% or more of regularly occupied spaces have multiple lines of sight to vision glazing in different directions at least 90 degrees apart
		2. 90% or more of regularly occupied spaces have views that include views of at least 2 of the following: 1) vegetation, 2) human activity, 3) objects at least 70 feet from the exterior of the glazing
		3. 90% or more of regularly occupied spaces have access to unobstructed views located within the distance of 3 times the head height of the vision glazing
		4. 90% or more of regularly occupied spaces have access to view with a view factor of 3 or greater

*counts toward SS Cr 1, Path 12 Other Quantifiable Performance

Appendix H
REFERENCED STANDARDS – LEED® CI v3.0

Sustainable Sites

SS Credit 1, Option 2: PATH 1	Brownfield Redevelopment	ASTM E1903-97, Phase II Environmental Site Analysis
		U.S. EPA Brownfields Definition
SS Credit 1, Option 2: PATH 3	Stormwater Design, Quality Control	Guidance Specifying Mgmt Measures for Sources of Non-Point Pollution in Coastal Waters (US EPA 840B92002)
		National Technical Information Service (PB93-234672)
SS Credit 1, Option 2: PATH 5	Heat Island Effect, Roof	ASTM E1980-01 – Standard Practice for Calculating SRI of Horizontal & Low-Sloped Opaque Surfaces
		ASTM E408-71 – Standard Test Methods for Total Normal Emittance of Surfaces Using Inspection-Meter Techniques
		ASTM E903-96 – Standard Test Methods for Solar Absorbance, Reflectance, & Transmittance of Materials using Integrating Spheres
SS Credit 1, Option 2: PATH 9	Innovative Wastewater Technologies	U.S. Energy Policy Act (EPAct) of 1992 and 2005
		International Association of Plumbing and Mechanical Officials Uniform Plumbing Code
		Publication IAPMO/ANSI UPC 1-2006
		International Code Council
SS Credit 1, Option 2: PATH 10	Water Use Reduction, 30% Reduction	U.S. Energy Policy Act (EPAct) of 1992 and 2005
		International Association of Plumbing and Mechanical Officials Uniform Plumbing Code, Section 402.0, Water-Conserving Fixtures and Fittings. Publication IAPMO/ANSI UPC 1-2006
		International Code Council, International Plumbing Code, Section 604, Design of Building Water Distribution System
SS Credit 1, Option 2: PATH 11	On-Site Renewable Energy	ASHRAE/IESNA 90.1-2007: Energy Standard for Buildings Except Low-Rise Residential
		Illuminating Engineers Society of North America

Water Efficiency

WE Prerequisite1	Water Use Reduction	U.S. Energy Policy Act (EPAct) of 1992 and 2005
		Uniform Plumbing Code, Section 402.0, Water-Conserving Fixtures and Fittings
		International Code Council, International Plumbing Code, Section 604, Design of Building Water Distribution System

Energy & Atmosphere

EA Prerequisite2	Minimum Energy Performance	ASHRAE/IESNA 90.1-2007, Energy Standard for Buildings Except Low-Rise Residential Buildings
EA Credit 1.1	Optimize Energy Performance: Lighting Power	ASHRAE/IESNA 90.1-2007, Energy Standard for Buildings Except Low-Rise Residential Buildings
EA Credit 1.3	Optimize Energy Performance: HVAC	ASHRAE/IESNA 90.1-2007, Energy Standard for Buildings Except Low-Rise Residential Buildings
		Advanced Buildings™ Core Performance™ Guide
EA Credit 1.4	Optimize Energy Performance: Equipment and Appliances	Energy Star® – Qualified Products
EA Credit 3	Measurement & Verification	IPMVP Volume 1, Concepts and Options for Determining Energy and Water Savings
EA Credit 4	Green Power	Center for Resource Solutions, Green-e Renewable Electricity Certification Program

Materials & Resources

MR Credit 4	Recycled Content	ISO 14021-1999, Environmental Labels and Declarations, Self-Declared Environmental Claims (Type II Environmental Labeling)
MR Credit 7	Certified Wood	Forest Stewardship Council's Principles & Criteria

Indoor Environmental Quality

EQ Prerequisite1	Minimum IAQ Performance	ASHRAE 62.1-2007 — Ventilation for Acceptable Indoor Air Quality
EQ Prerequisite2	Environmental Tobacco Smoke Control	ANSI/ASTM E779-03, Standard Test Method for Determining Air Leakage

		California Low Rise Residential Alternative Calculation Method Approval Manual, Home Energy Rating Systems (HERS) Required Verification and Diagnostic Testing
EQ Credit 1	Outside Air Delivery Monitoring	ASHRAE 62.1-2007 — Ventilation for Acceptable Indoor Air Quality
EQ Credit 2	Increased Ventilation	ASHRAE 62.1-2007 — Ventilation for Acceptable Indoor Air Quality
		The Carbon Trust Good Practice Guide 237 — Natural Ventilation in Non-Domestic Buildings, A Guide for Designers, Developers, & Owners
		CIBSE Applications Manual 10: 2005, Natural Ventilation in Non-Domestic Buildings
EQ Credit 3.1	Construction IAQ Management Plan: During Construction	ANSI/ASHRAE 52.2-1999 — Method of Testing General Ventilation Air-Cleaning Devices for Removal Efficiency by Particle Size
		IAQ Guidelines for Occupied Buildings Under Construction, SMACNA
EQ Credit 3.2	Construction IAQ Management Plan: Before Occupancy	U.S. EPA Compendium of Methods for the Determination of Air Pollutants in Indoor Air
EQ Credit 4.1	Low-Emitting Materials: Adhesives & Sealants	South Coast Air Quality Management District (SCAQMD) Rule #1168, VOC Limits
		Green Seal Standard GS-36 (Commercial Adhesives)
EQ Credit 4.2	Low-Emitting Materials: Paints & Coatings	South Coast Air Quality Management District (SCAQMD) Rule #1113, Architectural Coatings
		Green Seal Standard GC-03 (anti-corrosive & anti-rust paints)
		Green Seal Standard GS-11 (commercial flat & non-flat paints)
EQ Credit 4.3	Low-Emitting Materials: Flooring Systems	Carpet & Rug Institute (CRI) Green Label Plus Testing Program
		South Coast Air Quality Management District (SCAQMD) Rule #1168, VOC Limits
		South Coast Air Quality Management District (SCAQMD) Rule #1113, Architectural Coatings
		Floorscore Program
		State of California Specification Section 01350
		Environmental Technology Verification (ETV) Large Chamber Test Protocol for Measuring Emissions of VOCs and Aldehydes

EQ Credit 4.5	Low-Emitting Materials:	Greenguard™ Certification Program
	Systems Furniture & Seating	U.S. EPA Environmental Technology Verification (ETV) Large Chamber Test Protocol for Measuring Emissions of VOCs and Aldehydes (Sept 1999)
EQ Credit 5	Indoor Chemical & Pollutant Source Control	ANSI/ASHRAE 52.2-1999 — Method of Testing General Ventilation Air-Cleaning Devices for Removal Efficiency by Particle Size
EQ Credit 6.2	Controllability of Systems:	ASHRAE 62.1-2007 — Ventilation for Acceptable Indoor Air Quality
	Thermal Comfort	ASHRAE 55-2004 — Thermal Comfort Conditions for Human Occupancy
EQ Credit 7	Thermal Comfort	ASHRAE 55-2004 — Thermal Comfort Conditions for Human Occupancy
EQ Credit 8.1	Daylight and Views: Daylight	ASTM D1003-07E1, Standard Test Method for Haze and Luminous Transmittance of Transparent Plastics

Innovation in Design		
ID Credit 2	LEED Accredited Professional	LEED Accredited Professional, Green Building Certification Institute

Appendix I

ANSWERS TO QUIZ QUESTIONS

CHAPTER 2: SUSTAINABILITY AND LEED BASICS REVIEW

Q2.1. **E.** All of the four options listed are environmental benefits of green building design, construction, and operational efforts.

Q2.2. **B.** According to the EPA website, Americans typically spend about 90 percent of their time indoors.

Q2.3. **A.** Thirty-eight percent of energy in the United States is used for space heating, followed by lighting with 20 percent of energy usage.

Q2.4. **E.** All of the four options listed describe high-performance green building strategies.

Q2.5. **B.** Risk is collectively managed in an IPD. The risks and rewards are both shared in an IPD project.

Q2.6. **A.** Incorporating green building strategies and technologies is best started from the very beginning of the design process. Schematic design is the earliest phase of the design process, and, therefore, the correct answer.

Q2.7. **A.** It is important to remember that LCAs look at not only the present impacts and benefits during each phase of the process but future and potential impacts as well.

Q2.8. **B and C.** LCAs include the purchase price, installation, operation, maintenance, and upgrade costs for each technology and strategy proposed.

Q2.9. **A, D, and E.** Although it is strongly encouraged to begin the integrative design process and to incorporate green building technologies and strategies as early as possible in the design process, it is not intended to be an elaborate process. Value engineering should not be needed if the triple bottom line principles are applied.

Q2.10. **A, D and E.** Project teams working on a green building project should not experience increases to the project schedule or decreases in communication as a result of the integrated project delivery approach.

CHAPTER 3: THE LEED FOR COMMERCIAL INTERIORS RATING SYSTEM

Q3.1. **B and C.** Remember the point ranges for each level of certification.

Q3.2. **A and C.** Although the credits within the LEED rating systems are now weighted differently in the new version, the rating systems were not reorganized. The five main categories still exist, but now the rating systems are based on a 100-point scale. Each credit is now weighted in correlation to its impact on energy efficiency and CO_2 reductions.

Q3.3. **B.** Remember the LEED CS rating system is available to developers wishing to certify the base building, setting the pace for sustainable tenants.

Q3.4. **D.** USGBC consulted with NIST and the EPA's TRACI tool to determine the credit weightings by assessing carbon overlay.

Q3.5. **B.** Although the strategies listed will increase the first costs for a project, it is important to remember the life-cycle costs, including purchase price, installation, operation, maintenance, and upgrade costs.

Q3.6. **A.** Make sure to remember the point range scales of the LEED certification levels for the purposes of the exam.

Q3.7. **A and C.** CIRs are submitted to GBCI for review, electronically through LEED-Online. CIRs are limited to 600 words and should not be formatted as a letter. Since the CIR is submitted electronically through LEED-Online, the project and credit or prerequisite information is tracked; therefore, the CIR does not need to include this type of information. It is critical to remember that CIRs are submitted specific to one credit or prerequisite.

Q3.8. **A.** All tenant improvement projects seeking certification under the LEED CI rating system must include a minimum of 250 square feet of gross floor area per the MPRs listed on the USGBC website.

Q3.9. **B and E.** Be sure to remember each of the MPRs and how they pertain to each rating system, as posted on the GBCI website.

Q3.10. **E.** All projects seeking LEED certification must commit to sharing 5 years' worth of actual whole-project energy and water usage data with USGBC and/or GBCI.

Q3.11. **B and E.** Appeals are electronically submitted to GBCI through LEED-Online for a fee, within 25 business days after the final results from a design or construction certification review are posted to LEED-Online.

Q3.12. **C and E.** It is also important to remember, although the Regional Priority category is a new addition to the rating systems, no new prerequisites or credits were created to include within the new category. The Regional Priority category offers bonus points for achieving existing LEED credits detailed in the other categories.

Q3.13. **B.** Although design reviews can be beneficial, points are not awarded until final review after construction.

Q3.14. **D and E.** Registering with GBCI indicates a project is seeking LEED certification. GBCI responds to any CIRs. Project registration can be completed at any time, although one is strongly encouraged to do so as early as possible. GBCI does not grant the award of any points regardless when registration occurs. Registration will, however, grant the project team access to a LEED-Online site specific for the project, but it does not include any free submissions of CIRs.

Q3.15. **A and C.** Although registering a project requires some information, including contact information, project location, and indication of compliance with MPRs, a team must submit all credit forms for all prerequisites and attempted credits during the certification application. Required supplemental documentation, such as plans and calculations, must be uploaded as well.

Q3.16. **C.** The earliest construction prerequisites and credits can be submitted, along with design prerequisites and credits, for certification review is after substantial completion.

Q3.17. **C.** It is important to remember that points are awarded only once the project team submits for construction review, not at the design phase certification review. Design-side review is optional and, therefore, not required.

Q3.18. **B.** Project teams have 25 business days to issue an appeal to GBCI after receiving the final review comments.

Q3.19. **B, C, and F.** CIRs can be submitted through LEED-Online any time after registration. CIRs specifically address one MPR, prerequisite, or credit. Although the project administrator submits the CIR, the GBCI response is viewable by all team members invited to the LEED-Online site for the project. CIRs are project specific and, therefore, CIR responses will no longer be published to a database as they once were.

Q3.20. **A.** GBCI is responsible for the appeals process. It is best to remember USGBC as an education provider for the LEED rating systems they create, and GBCI as being responsible for the professional accreditation and project certification processes.

CHAPTER 4: SUSTAINABLE SITES

Q4.1. **B.** Where a project is located and how it is developed can have multiple impacts on the ecosystem and water resources required during the life of a building.

Q4.2. **B.** In order to determine if a portion of land is a brownfield site, an examination is done via the ASTM E1903-97 Phase II Environmental Site Assessment.

Q4.3. **A, B, and D.** You may find questions that you are not familiar with but use these to learn more content in a different manner. It also prepares you for the real exam as new questions are consistently presented.

Q4.4. **A and C.** Be sure to remember the difference between the three types of remediation strategies.

Q4.5. **C.** The key to reducing heat island effects is to avoid implementing materials that will absorb and retain heat. Deciduous trees lose their leaves and, therefore, are not the best decision. Teams should look to trees that can provide consistent shade within 5 years after planting. Xeriscaping to reduce evaporation and increasing impervious surfaces to recharge groundwater are great strategies for sustainable site design, but they do not help to reduce heat island effects. In turn, they reap the benefits of reduced heat island effects. Implementing paving and roofing products with a higher albedo, or SRI, is, therefore, the best answer to reduce heat island effects.

Q4.6. **A, B, D, and F.** Make a flashcard to remember these types of benefits.

Q4.7. **D.** Emissivity is the ability of a material surface to give up heat in the form of radiation. It may be helpful to remember that emittance is the opposite of reflectivity. Infrared reflectivity applies to low-emissivity materials. Therefore, these materials reflect the majority of long-wave radiation and emit very little, such as metals or special metallic coatings. High-emissivity surfaces, such as painted building materials, absorb a majority of long-wave radiation as opposed to reflecting it, and emit infrared or long-wave radiation more willingly.

Q4.8. **A, B, and C.** The LEED CI rating system recommend to combine the following three strategies to reduce heat island effects: provide shade (within five years of occupancy), install paving materials with an SRI of at least 29, and implement an open-grid pavement system (less than 50 percent impervious).

Q4.9. **C.** Remember, some material in this book can be presented for the first time in the format of a question. Use these opportunities to test your knowledge, and if you are not familiar with the content, be sure to make a flashcard or make a note on your cheat sheet in order to remember the information.

Q4.10. **C.** Subtracting 180,000 from 360,790 results in 180,790 gallons for the design case water demand for irrigation, with the baseline case requiring 2,027,825 gallons. Dividing the two values and subtracting from 1 results in 0.911, then multiplying by 100 is required to get to the answer of 91 percent savings.

Q4.11. **A and C.** Be sure to use your flashcards or your cheat sheet to remember the factors to determine irrigation efficiency. The process of elimination might work, too.

Q4.12. **B, C, and E.** It might not be feasible to make notes to remember all of the documentation requirements for each prerequisite and credit for the exam. If you know the strategies to comply and the terminology, you will be able to decipher the applicable answers.

Q4.13. **A.** The process of elimination might be helpful to use to deduce the most appropriate answer.

Q4.14. **A, B, and C.** According to the *Green Building and LEED Core Concepts Guide,* transportation is most affected by location, vehicle technology, fuel, and human behavior.

Q4.15. **A, C, and D.** If a LEED project's site does not offer mass transit accessibility and is, therefore, dependent on commuting by car, it is best to encourage the occupants to carpool or incorporate conveniences within the building or on-site.

Q4.16. **A, B, and C.** Selecting a site near public transportation, limiting parking, and encouraging carpooling are all strategies to consider when working on a project seeking LEED certification. It is always best to redevelop a previously developed site, avoiding greenfield sites.

Q4.17. **B, E, and F.** Passive solar design features and off-site strategies do not contribute to earning the On-Site Renewable Energy credit. Ground-source heat pumps do not qualify either, as they require power to for the pump to function.

Q4.18. **C and E.** Be sure to remember projects that operate 24 hours a day are required to comply with Option 2 for this credit.

Q4.19. **B.** Be sure to use the study worksheets at the end of each chapter to help remember the details of each prerequisite and credit.

Q4.20. **A, C, and D.** Impervious asphalt does not allow rainwater to percolate through and, therefore, allows stormwater to leave the site, carrying pollutants and debris, heading to storm sewers and nearby bodies of water.

Q4.21. **C and E.** Project teams can evaluate location and site-specific information prior to the beginning of designing a structure and the site to determine the efficiencies of strategies and technologies for a green building project. These issues include the availability of mass transit and public transportation, and brownfield redevelopment. Strategies to reduce heat island effects, provisions for preferred parking, and technologies to reduce water use should all be addressed by the building owner prior to signing a lease or provisions will need to be made in order to pursue points under SS Credit 1.

Q4.22. **D.** Access to 10 basic services, such as banks, post offices, grocery stores, schools, restaurants, fire stations, hardware stores, pharmacies, libraries, theaters, museums, and fitness centers, is required to comply with LEED.

Q4.23. **D and E.** Use your flashcards to remember these types of details. You most likely will not see an exam question asking about what an acronym might stand for, but it will be helpful to be familiar with the acronyms, as they might be part of an answer selection.

Q4.24. **D and E.** Selecting products with the highest SRI values is best suited for compliance with LEED.

Q4.25. **A and B.** Learning the strategies to comply with each of the credits and prerequisites is key for the exam. Try to remember the intent behind each to help decipher the appropriate strategy or technology.

Q4.26. **B and E.** Trust your memory and your logical senses! Selecting a greenfield site is most likely not going to be rewarded within LEED, such as offering a point opportunity. Remember, a graded site is a previously developed site. Last, 10 basic services are required not 5.

Q4.27. **A, B, and D.** Unfortunately, for the purposes of LEED, carpooling and vanpooling do not count as public transportation strategies. For a citywide approach to ride sharing, try doing a web search for "DC slug lines."

Q4.28. **B.** In order to qualify for this credit, 50 percent of the roof surface must comply by the means of a high-SRI roofing material and/or a green roof. Photovoltaic panels and mechanical equipment square footage is subtracted from the compliant roof surface calculations. Therefore, 50,000 – 15,000 – 5,000 = 30,000 eligible roof surface, where 50 percent would need to comply with the minimum SRI level.

Q4.29. **A.** First, determine the FTE, whereby full-time employees have a value of 1 and part-time employees working an average of four hours per day have half value; therefore the FTE = 700. For bike parking, 5 percent of 700 = 35 and for showers, 0.5 percent of 700 = 3.5. For the purposes of LEED, it is appropriate to round up values for compliance.

Q4.30. **C.** Remember to use your cheat sheet for details difficult to remember!

CHAPTER 5: WATER EFFICIENCY

Q5.1. **B and D.** Sometimes the process of elimination helps to determine the correct answers. Although captured rainwater is used for custodial uses, cleaning dishes and clothes is best with potable water sources.

Q5.2. **B and F.** Remember flow fixture and flush fixture types for the exam.

Q5.3. **C.** Use your flashcards to remember these types of details.

Q5.4. **A, D, and E.** Remember the components addressed by EPAct.

Q5.5. **E.** Don't forget about Exemplary Performance under the ID category and Regional Priority!

CHAPTER 6: ENERGY AND ATMOSPHERE

Q6.1. **C.** Use your flashcards and cheat sheet to remember these type of credit details.

Q6.2. **D.** Remember to use your cheat sheet for details difficult to remember!

Q6.3. **D.** Use your flashcards to quiz yourself on these types of details.

Q6.4. **C.** Be sure to remember the reference standards for the purposes of the exam.

Q6.5. **E.** Refer to your flashcards to remember the different reference standards for the two compliance paths for this credit.

Q6.6. **A and B.** Answer options C, D, and E indicate the incorrect percentage and distance thresholds to comply. This leaves options A and B should you have not known they were correct, thus utilizing the process of elimination. In terms of lighting controls, using continuously dimmed controls is best, but step dimming or even a regular on/off switch will also comply with the credit requirements.

Q6.7. **C.** Use your flashcards to quiz yourself on these types of details or make a note on your cheat sheet.

Q6.8. **A.** Remember, it is advisable to incorporate commissioning tasks, such as the OPR and BOD, as early as possible in the design process.

Q6.9. **B and D.** Remember, commissioning agents should be independent third parties to perform their responsibilities for the owner. A CxA is responsible for minimizing design flaws and assessing the installation, calibration, and performance for the main building systems.

Q6.10. **B and E.** Be sure to use the study worksheets at the end of each chapter to help remember the details of each prerequisite and credit.

Q6.11. **C.** Consult Appendix H for a list of all of the referenced standards and use the study worksheets at the end of each chapter to help remember the details of each prerequisite and credit.

APPENDIX I **285**

Q6.12. **A.** LEED CI only addresses one type of refrigerant for compliance; be sure to remember it.

Q6.13. **B, D, and E.** Process energy uses include computers, office equipment, kitchen refrigeration and cooking, washing and drying machines, and elevators and escalators. Miscellaneous items, such as a waterfall pumps and lighting that is exempt from lighting power allowance calculations, such as lighting integrated into equipment, are also categorized as process energy uses.

Q6.14. **A and D.** Refrigerants do not apply to boilers, fan motors, or variable-frequency drives. To meet the requirements of the prerequisite, the HVAC systems within the tenant's scope of work is required to comply not base building systems.

Q6.15. **B, E, and F.** It is critical to remember the details about commissioning for the purposes of the exam.

Q6.16. **C.** Remember each of the referenced standards and what each applies to. Remember to think "energy" every time you read "ASHRAE 90.1"!

Q6.17. **B, D, and E.** Regulated energy uses include lighting, HVAC, and domestic and space heating for service water.

Q6.18. **C.** Be sure to use the study worksheets at the end of the chapter to remember the sequence of equations.

Q6.19. **C.** Green power should be remembered with off-site renewable energy, as green power is purchased and not installed.

Q6.20. **A and B.** Remember Table 6.2! There are 11 tasks to comply with the Enhanced Commissioning credit, which is 5 more tasks than what is required by the prerequisite.

Q6.21. **B.** Use the process of elimination to deduce the correct answer.

Q6.22. **C and D.** Remember no prerequisites are eligible for exemplary performance. Surpassing the minimum thresholds are addressed in the credits.

Q6.23. **B.** Remember if the local ordinances or codes are more strict than the reference standard, they can be replace the standard.

Q6.24. **B and C.** Efficacy relates to how well a light fixture produces light, so higher is better. Utilizing fixtures with a high reflectance value results in distributing more light; therefore, fewer fixtures are required.

Q6.25. **C and D.** Be sure to use your flashcards to remember the requirement details of each prerequisite and credit.

CHAPTER 7: MATERIALS AND RESOURCES

Q7.1. **D.** First, determine the eligible items based on the requirements of the credit. The doors are not eligible, as they are not repurposed; therefore, the total amount reused is $120,000 of the total $2 million, for 17 percent compliance. The percentage threshold for CI projects = 15 percent.

Q7.2. **B and D.** Remember the differences among the intentions and requirements of MR Credit 4: Recycled Content and Credit 3: Material Reuse suite. Whenever you see "nearby," try to remember the possible connection to MR Credit 5: Regional Materials.

Q7.3. **B, D, and E.** Be sure to refer to Appendix G for a list of the exemplary performance opportunities.

Q7.4. **A and E.** Window assemblies are excluded, as it is assumed that the window is a part of the base building shell. Hazardous materials are excluded for obvious reasons, and structural materials are not addressed in the rating system.

Q7.5. **A and B.** Use your flashcards to quiz yourself on these types of details or make a note on your cheat sheet.

Q7.6. **C, D, and E.** Remember, preconsumer recycled content refers to scrap and trim material generated from the manufacturing process, but it does not enter into the consumer cycle of goods. Preconsumer recycled materials are used to manufacture a different product than what it was originally intended for.

Q7.7. **A and D.** Regional materials, salvaged materials, and rapidly renewable materials are calculated as a percentage of the total material cost for a project for the purposes of LEED.

Q7.8. **B.** Rapidly renewable products can be grown or raised in 10 years or less.

Q7.9. **B and D.** For the purposes of LEED, FSC wood products are calculated as a percentage of the total cost of wood products purchased for a specific project. Chain-of-custody documentation should be tracked and the certification number entered into the credit forms for proof of compliance.

Q7.10. **D.** This is a characteristic unique to the LEED CI rating system.

Q7.11. **C, E, and F.** Use your flashcards to quiz yourself on these types of details or make a note on your cheat sheet.

Q7.12. **A and B.** Be sure to refer to Appendix J for a list of all acronyms and abbreviations.

Q7.13. **B.** Landfills produce methane, a powerful greenhouse gas. Although methane can be captured and burned to generate energy, if it is emitted, it is harmful to the environment.

Q7.14. **B and C.** Cradle-to-cradle products can be recycled, while cradle-to-grave materials are landfilled. Products with either or both preconsumer and postconsumer recycled content can contribute to earning the LEED credit.

Q7.15. **B and C.** Stormwater management plans are typically the responsibility of the civil engineer, while the energy modeling calculations are typically provided by the mechanical engineer.

Q7.16. **B, D, and E.** Although evaluating all of the vendor's procurement policies to ensure that sustainable purchasing procedures are in place could possibly contribute to earning an Innovation in Design credit, it is not required for compliance with any of the MR credits. The CEO's automobile choice is also not evaluated or assessed for LEED compliance.

Q7.17. **D.** FSC credit compliance requires the completion of a credit form, including invoice amounts and chain-of-custody certification numbers. Remember, all documentation is submitted for review via LEED-Online for all projects.

Q7.18. **D.** Waste is calculated in tonnage for the purposes of LEED documentation.

Q7.19. **C and D.** Use your flashcards to quiz yourself on these types of details or make a note on your cheat sheet.

Q7.20. **B.** It is critical to know what factors are appropriate to be included in the calculation. The 20 tons of wood, plastic, and metal and 45 tons of concrete = 65 tons diverted. The 20 tons landfilled plus the 15 tons sent to an incineration plant = 35 tons not diverted. 65 + 35 = 100 total tons, as the 8 tons are not eligible to be included in the calculations.

CHAPTER 8: INDOOR ENVIRONMENTAL QUALITY

Q8.1. **A and C.** Be sure to remember which credits are eligible for a design side review. Knowing what the documentation requirements are can help with the process of elimination for answer selections.

Q8.2. **C.** Use your flashcards to quiz yourself on these types of details or make a note on your cheat sheet.

Q8.3. **D.** Be sure to use the study worksheets at the end of each chapter to help remember the details of each prerequisite and credit.

Q8.4. **B.** Use your flashcards to quiz yourself on these types of details or make a note on your cheat sheet.

Q8.5. **E.** Use your flashcards to quiz yourself on these types of details or make a note on your cheat sheet.

Q8.6. **D.** LEED requires a minimum of MERV 8 filters to be installed for compliance.

Q8.7. **A and F.** Be sure to review Appendix H for the referenced standards.

Q8.8. **B, E, and F.** Remember, the specifics of the referenced standards and what they apply to. Phenol-formaldehyde and urea-formaldehyde relate to resin-manufactured building materials. Knowing the requirements of each of the materials would have helped to eliminate answer option C. Option D is eliminated, as Green Spec is not currently included in the referenced standards addressed in the LEED rating systems.

Q8.9. **D.** Remember the specifics of the referenced standards and what they apply to.

Q8.10. **D.** Be sure to make a flashcard and make a note on your cheat sheet to remember the SMACNA standards.

Q8.11. **D, E, and F.** Remember only furniture associated with workstations are required to comply in order to earn this credit. This would eliminate occasional furniture for reception areas and break rooms and desks for private offices that are not part of a larger system and gathering.

Q8.12. **A, C, and E.** Use your flashcards to help remember which reference standards are associated with which credit. Appendix H summarizes all of the reference standards of the CI rating system in one chart.

Q8.13. **A, C, and D.** ASHRAE 90.1 = Energy! ENERGY STAR applies to energy-efficient appliances, products, and buildings.

Q8.14. **A, C, and D.** Read through each answer selection carefully or try hiding the answers and first derive your own before revealing the selections.

Q8.15. **A and E.** Use your flashcards to quiz yourself on these types of details or make a note on your cheat sheet.

Q8.16. **D.** Use your flashcards to quiz yourself on these types of details or make a note on your cheat sheet.

Q8.17. **B and D.** ASHRAE 55 defines the three environmental components that impact thermal comfort, including humidity, air speed, and temperature.

Q8.18. **D.** Remember to read questions and answer options carefully to eliminate the incorrect answers and to depict the correct answer.

Q8.19. **C.** Use your flashcards to quiz yourself on these types of details or make a note on your cheat sheet.

Q8.20. **D.** Think about the logic behind this requirement. Is it really feasible to provide any higher of a percentage?

Q8.21. **D.** Glare control is required for each of the compliance paths as it is the most commonly forgotten element of most daylight strategies.

Q8.22. **A, B, and C.** Learning the strategies to comply with each of the credits and prerequisites is key for the exam. Try to remember the intent behind each to help decipher the appropriate strategy or technology.

Q8.23. **B, C, and E.** Remember, some material in this book can be presented for the first time in the format of a question. Use these opportunities to test your knowledge, and if you are not familiar with the content, be sure to make a flashcard or make a note on your cheat sheet in order to remember the information.

Q8.24. **E.** Use your flashcards to quiz yourself on these types of details or make a note on your cheat sheet.

Q8.25. **D.** Create a flashcard to remember this technical detail.

Q8.26. **C, D, and E.** The square footage indication is irrelevant. If any part of the envelope is being enhanced, typically EA Credit 1 is relevant. Remember to look for terms such as *nearby* and *adjacent* if MR Credit 5 is proposed as an answer choice. The Material Reuse credit comes into play with the salvaged wood floors. There was no mention of incorporating certified wood or low-emitting materials, thus eliminating answer options A and B.

Q8.27. **D.** 62 IAQ, IAQ 62!

Q8.28. **A and D.** Remember the BAIT Tip to point out the trade-offs and synergies of increased ventilation strategies: better IAQ but reduced energy efficiency for mechanical systems to condition outside air.

Q8.29. **A and E.** Project specifications will give the contractor direction on how to comply with the IAQ credit requirements such as MERV filters, flush-out, and low-emitting materials.

Q8.30. **A and C.** Remember ASHRAE 55 = thermal comfort and 62 = IAQ, as both relate to ventilation system design.

Q8.31. **A and C.** Proximity to a shopping mall may increase satisfaction because of its convenience but not necessarily increase production as related to work. Carpooling and recycling are benefits to the environment and operations, not necessarily related to productivity or satisfaction.

Q8.32. **A and C.** Be sure to remember which credits are eligible for a design-side review. Knowing what the documentation requirements are can help with the process of elimination for answer selections.

Q8.33. **B, D, and E.** Use your flashcards to quiz yourself on these types of details or make a note on your cheat sheet.

Q8.34. **A, C, and E.** Use your flashcards to quiz yourself on these types of details or make a note on your cheat sheet.

Q8.35. **C.** Thermal comfort is not related to this credit. Although it will improve air quality, the occupant's ability to control their thermal comfort is not addressed.

Q8.36. **B.** 14,000 cu ft of outside is required per square foot to comply with the credit.

CHAPTER 9: INNOVATION IN DESIGN AND REGIONAL PRIORITY

Q9.1. **C.** Earning exemplary performance is credit specific, so be aware of statements such as "regardless of which credit is being exceeded."

Q9.2. **C and D.** If unclear about ID credits, be sure to read through the Guidance on Innovation and Design (ID) credits on the GBCI website at www.gbci.org/ShowFile.aspx?DocumentID=3594.

Q9.3. **A.** A new green building project can earn the LEED NC certification and then earn the LEED EBOM certification during operations or a LEED Core & Shell building can be built and then earn multiple LEED CI certifications.

Q9.4. **A, C, and E.** Remember, the Regional Priority category is new but does not include any new credits. RPCs are earned by achieving existing LEED credits from other categories. Although earning a maximum of four RPCs is allowed, there are six opportunities available from which to choose.

Q9.5. **D.** Although earning a maximum of four RPCs is allowed, there are six opportunities available from which to choose for each zip code.

Q9.6. **A.** Regardless of how many LEED APs are on a project, only 1 point can be earned.

Q9.7. **D.** Remember to study Appendix G to learn all of the exemplary performance opportunities within the LEED CI rating system.

Q9.8. **D, E, and F.** This one was tricky, as the four measures were mentioned briefly in the previous chapter.

Q9.9. **A.** Use those flashcards! Remember MR Credit 1.2 uses square footage and area calculations for documenting compliance.

Q9.10. **A.** This is a unique characteristic to the LEED CI rating system where the SS category does not have a prerequisite. Remember, although a project team may not pursue credits within each of the categories, all of the prerequisites must be met.

Q9.11. **B.** Be sure to review Appendix H for the referenced standards.

Q9.12. **C.** Be sure to use the 8 kw/sf/yr default value to calculate electricity consumption for the purposes of the exam. Don't forget the term for the purchase agreement must be 2 years to comply with the credit. $44,000 \times 8 \times 2 = 704,000$ kwh

Q9.13. **A.** Be sure to make note a steep-sloped roof requires a roofing material with a minimum SRI value of 29 while a low-sloped roof requires a roofing material with a minimum SRI value of 78. Be sure to also remember 2:12 as the threshold distinguishing a roof from low-sloped to steep-sloped. These are great cheat sheet details.

Q9.14. **A, B, C, and E.** The process of elimination might make this one easier to answer correctly.

Appendix J

ABBREVIATIONS AND ACRONYMS

4-PCH	4-phenylcyclohexene
AFV	alternative-fuel vehicle
AIA	American Institute of Architects
ANSI	American National Standards Institute
AP	LEED Accredited Professional
ASHRAE	American Society of Heating, Refrigerating, and Air-Conditioning Engineers
ASTM	American Society for Testing and Materials
BAS	building automation system
BD+C	Building Design and Construction (LEED AP credential and also a reference guide)
BEES	Building for Environmental and Economic Sustainability software by NIST
BIFMA	Business and Institutional Furniture Manufacturer's Association
BIPV	building integrated photovoltaics
BIM	building information modeling
BMP	best management practice
BOD	basis of design
BOMA	Building Owners and Managers Association
CAE	combined annual efficiency
CBECS	Commercial Building Energy Consumption Survey (by DOE)
CDs	construction documents
CDL	construction, demolition, and land clearing
CE	controller efficiency
CFA	conditioned floor area
CFC	chlorofluorocarbon
CFL	compact fluorescent light
CFM	cubic feet per minute
CFR	U.S. Code of Federal Regulations
CI	Commercial Interiors (LEED CI rating system)
CIBSE	Chartered Institution of Building Services Engineers
CIR	credit interpretation request
CMP	Credentialing Maintenance Program
CO	carbon monoxide
CO_2	carbon dioxide
COC	chain of custody
COP	coefficient of performance
CRI	Carpet and Rug Institute
CS	Core & Shell (LEED CS rating system)
CSI	Construction Specifications Institute
CWMP	construction waste management plan
Cx	commissioning
CxA	commissioning agent or authority
dBA	A-weighted decibel
DHW	domestic hot water
DOE	U.S. Department of Energy

EA	Energy & Atmosphere category
EBOM	Existing Buildings: Operations & Maintenance (LEED EBOM rating system)
ECB	energy cost budget
ECM	energy conservation measure
EER	energy efficiency rating
EERE	U.S. Office of Energy Efficiency and Renewable Energy
EF	energy factor
EPA	U.S. Environmental Protection Agency
EPAct	U.S. Energy Policy Act of 1992 or 2005
EPEAT	electronic product environmental assessment tools
EPP	environmentally preferable purchasing
ESA	environmental site assessment
ESC	erosion and sedimentation control
ET	evapotranspiration
ETS	environmental tobacco smoke
ETV	Environmental Technology Verification
EQ	Indoor Environmental Quality category
fc	footcandle
FEMA	U.S. Federal Emergency Management Agency
FF&E	fixtures, furnishings, and equipment
FSC	Forest Stewardship Council
FTE	full-time equivalent
GBCI	Green Buildings Certification Institute
GBOM	Green Buildings Operations + Maintenance reference guide
GF	glazing factor
GHG	greenhouse gas
GPF	gallons per flush
g/L	grams per Liter
GPM	gallons per minute
GWP	global warming potential
HCFC	hydrochlorofluorocarbon
HEPA	high-efficiency particle absorbing
HERS	Home Energy Rating Standards
HET	high-efficiency toilet
HFC	hydrofluorocarbon
HVAC	heating, ventilation, and air conditioning
HVAC&R	heating, ventilation, air conditioning, and refrigeration
IAQ	indoor air quality
ICF	insulated concrete form
ID	Innovation & Design category
ID+C	Interior Design + Construction (LEED AP credential and also a reference guide)
IE	irrigation efficiency
IEQ	Indoor Environmental Quality category
IESNA	Illuminating Engineering Society of North America
IPD	integrated project delivery
IPM	integrated pest management
IPMVP	International Performance Measurement and Verification Protocol
ISO	International Organization for Standardization
KW	kilowatt
KWh	kilowatt per hour

LCA	life-cycle assessment/analysis
LCC	life-cycle cost
LCGWP	life-cycle global warming potential
LCODP	life-cycle ozone depletion potential
LED	light-emitting diode
LEED	Leadership in Energy and Environmental Design
LPD	lighting power density
MDF	medium-density fiberboard
MERV	minimum efficiency reporting value
MPR	Minimum Program Requirement
MR	Materials & Resources category
MSDS	material safety data sheet
M&V	measurement and verification
NBI	New Building Institute
NC	New Construction (LEED NC rating system)
ND	Neighborhood Development (LEED ND rating system)
NH_3	ammonia
NIST	National Institute of Standards and Technology
NRC	noise reduction coefficient
ODP	ozone-depleting potential
O&M	operations and maintenance
O+M	Operations + Maintenance (LEED AP credential)
OPR	owner's program requirements
OSB	oriented strand board
PV	photovoltaic
PVC	polyvinyl chloride
REC	renewable energy certification
RESNET	Residential Energy Services Network
RFC	request for clarification
RFI	request for information
RFP	request for proposal
RP	Regional Priority category
RT	reverberation time
SCAQMD	South Coast Air Quality Management District
SCS	Scientific Certification Systems
SEER	seasonal energy efficiency rating
SHGC	solar heat gain coefficient
SIP	structural insulated panels
SMACNA	Sheet Metal and Air-Conditioning Contractor's Association
SS	Sustainable Sites category
SRI	solar reflective index
STC	standard transmission class
TAG	Technical Advisory Group
TASC	Technical Advisory Subcommittee
TP	total phosphorus
TRACI	Tool for the Reduction and Assessment of Chemical and Other Environmental Impacts
TSS	total suspended solids
Tvis	visible transmittance
UL	Underwriter's Laboratory
USGBC	U.S. Green Building Council

VLT	visible light transmittance
VOC	volatile organic compound
WE	Water Efficiency category
WF	water factor
WFR	window-to-floor ratio
WWR	window-to-wall ratio
ZEV	zero emission vehicle

Appendix K

SAMPLE CREDIT[1]

EQ CREDIT 4.4: LOW-EMITTING MATERIALS—COMPOSITE WOOD AND AGRIFIBER PRODUCTS

1 Point

Intent

To reduce the quantity of indoor air contaminants that are odorous, irritating, and/or harmful to the comfort and well-being of installers and occupants.

Requirements

Composite wood and agrifiber products used on the interior of the building (i.e., inside the weatherproofing system) must contain no added urea-formaldehyde resins. Laminate adhesives used to fabricate on-site and shop-applied composite wood and agrifiber assemblies must not contain added urea-formaldehyde resins.

Composite wood and agrifiber products are defined as: particleboard, medium-density fiberboard (MDF), plywood, wheatboard, strawboard, panel substrates and door cores. Materials considered fixtures, furniture, and equipment (FF&E) are not considered base building elements and are not included.

Products covered by EQ Credit 4.5, Low-Emitting Materials, System Furniture and Seating are excluded from these requirements.

Potential Technologies & Strategies

Specify wood and agrifiber products that contain no added urea-formaldehyde resins. Specify laminating adhesives for field and shop applied assemblies, including adhesives and veneers that contain no urea-formaldehyde. Review product cut sheets, material safety data (MSD) sheets, signed attestations, or other official literature from the manufacturer.

Credits

CHAPTER 1

1. USGBC website, www.usgbc.org/DisplayPage.aspx?CMSPageID=19.
2. GBCI. *LEED AP Interior Design + Construction Candidate Handbook*, March 2011 (2008), p. 7.

CHAPTER 2

1. Wikipedia website, http://en.wikipedia.org/wiki/Sustainability.
2. United Nations General Assembly, *Report of the World Commission on Environment and Development: Our Common Future* (1987). Transmitted to the General Assembly as an Annex to document A/42/427—Development and International Co-operation: Environment. Retrieved on February 1, 2009. www.un.org/documents/ga/res/42/ares42-187.htm.
3. USGBC website, www.usgbc.org/DisplayPage.aspx?CMSPageID=1718.
4. GSA Public Buildings Service, "Assessing Green Building Performance: A Post Occupancy Evaluation of 12 GSA Buildings" (2008).
5. Environmental Protection Agency. *The Total Exposure Assessment Methodology (TEAM) Study* (1987).
6. See note 1.
7. Heschong Mahone Group, "Daylighting in Schools: An Investigation into the Relationship Between Daylighting and Human Performance" (1999).
8. USGBC website, www.usgbc.org/DisplayPage.aspx?CMSPageID=1718.
9. USGBC website, www.usgbc.org/DisplayPage.aspx?CMSPageID=124.
10. USGBC website, www.usgbc.org/DisplayPage.aspx?CMSPageID=124.
11. GBCI website, www.gbci.org/org-nav/about-gbci/about-gbci.aspx.
12. GBCI website. "Disciplinary and Exam Appeals Policy", www.gbci.org/Files/Disc_ExamAppeals_Policy.pdf.

CHAPTER 3

1. USGBC, *LEED Reference Guide for Green Interior Design and Construction*, (2009), p. xiv.
2. USGBC website, www.usgbc.org/ShowFile.aspx?DocumentID=6715.
3. Ibid.

CHAPTER 4

1. EPA website, http://epa.gov/brownfields/about.htm.
2. EPA website, http://epa.gov/brownfields/.
3. USGBC. *LEED Reference Guide for Green Interior Design and Construction* (2008), p. 25.
4. Ibid., p. 34.
5. Ibid., p. 45.
6. USGBC, *Green Building and LEED Core Concepts Guide*, p. 27.

7. Ibid.
8. Wikipedia website. http://en.wikipedia.org/wiki/Vehicle_Emissions.
9. USGBC. *Green Building and LEED Core Concepts Guide*, 1st edition (2009), p. 28.
10. "IPCC AR4 SYR Appendix Glossary" (PDF). Retrieved on April 20, 2010.

CHAPTER 6

1. USGBC website, www.usgbc.org/DisplayPage.aspx?CMSPageID=1718.
2. USGBC. *LEED Reference Guide for Green Interior Design and Construction* (2008), p. 155.
3. Ibid., p. 156.
4. Ibid.
5. Ibid., p. 161.
6. USGBC, *Green Building and LEED Core Concepts Guide*, p. 51.

CHAPTER 7

1. County of San Mateo, California. *San Mateo Countywide Guide Sustainable Buildings* (2004), p. 3. www.recycleworks.org/pdf/GB-guide-2-23.pdf. Accessed December 2009.
2. Department of Natural Resources, Northeast Region. "Building Green at DNR—Northeast Region Headquarters Construction Waste & Recycling."
3. USGBC, *LEED Reference Guide for Green Interior Design and Construction* (2009), p. 430.
4. Ibid., p. 444.
5. Ibid., pp. xi and 207.
6. USGBC, *Green Building and LEED Core Concepts Guide*, p. 55.
7. USGBC, *LEED Reference Guide for Green Interior Design and Construction* (2009), p. 207.

CHAPTER 8

1. Environmental Protection Agency. The Total Exposure Assessment Methodology (TEAM) Study (1987).
2. USGBC, *Green Building and LEED Core Concepts Guide,* p. 59.
3. USGBC, *LEED Reference Guide for Green Interior Design and Construction* (2009), p. 438.
4. USGBC, *Green Building and LEED Core Concepts Guide,* p. 60.
5. USGBC, *LEED Reference Guide for Green Interior Design and Construction* (2009), p. 312.
6. Ibid., p. 429.
7. USGBC, *Green Building and LEED Core Concepts Guide,* p. 60.

APPENDIX F

1. USGBC website. www.usgbc.org/DisplayPage.aspx?CMSPageID=145.

APPENDIX K

1. USGBC website. www.usgbc.org/DisplayPage.aspx?CMSPageID=145.

Index

The letter t following a page number indicates a table.

4-PCH. *See* 4-phenylcyclohexene
4-phenylcyclohexene, 192. *See also* Contaminants

A

Adaptively reused, 149
Adaptive plantings, 48–51
Advanced Buildings Core Performance Guide, 117
Albedo, 45–46
Alternative transportation, 39t, 60–63, 94–99
American National Standards Institute (ANSI), 290
American Society for Testing and Materials (ASTM), 290
American Society of Heating, Refrigerating and Air-Conditioning Engineers (ASHRAE), 115, 187
ANSI. *See* American National Standards Institute
 /ASTM E779-03, Standard Test Method for Determining Air Leakage, 198
 /BIFMA
  M7.1 2007, 196
  x7.1-2007, 196
 /ISO/IEC 17024, 5
Aquifer, 42, 53
Architect, 13–15
ASHRAE. *See* American Society of Heating, Refrigerating and Air-Conditioning Engineers Standard
 55-2004, 199–201, 207
 62.1-2007, Ventilation for Acceptable Indoor Air Quality, 187, 200, 204, 207
 90.1-2007, 115, 118, 126

ASTM. *See* American Society for Testing and Materials
 E1903-97 Phase II Environmental Site Assessment, 57

B

BAS. *See* Building automation system
Baseline:
 energy usage, for, 115
 water usage, for, 102
Basis of design (BOD), 121, 123
Bernheim, Anthony, 257
Bicycle storage and changing rooms, 39t, 61–62, 62t, 64–65, 67, 96–97
BIM. *See* Building Information Modeling
Biodiversity, 42
Bioswales, 43
Blackwater, 52
BOD. *See* Basis of design
Boundary line, 28–29
Brownfield, 41, 57
 redevelopment, 41, 70–71
Building:
 automation system (BAS), 188
 density, 55
 envelope, 112
 footprint, 28–29
 Information Modeling (BIM), 16, 115
 operations, 9
 orientation, 110–111
 reuse, 149t, 150, 152, 168–169
 systems design, 9, 121
Buildings, conventionally designed and built, 10–11, 13–16
Business:
 community connectivity and, 55
 Indoor Environmental Quality impacts on
  absenteeism, reducing, 185

 employee satisfaction, 185, 202, 207
 productivity, 202, 207

C

C2C. *See* Cradle-to-cradle certified products
Carbon dioxide (CO_2):
 emissions, 39, 109, 159, 190 (*See also* Contaminants)
 reduction strategies, 24
 sensors, 188
Carbon monoxide (CO), 192
Carbon Trust's *Good Practice Guide 237,* 188
Carpet and Rug Institute (CRI), 190
 Green Label program, 194, 196
Carpools/vanpools, 62–64
CDs. *See* Construction documents
Certified wood. *See* Wood: Certified
CFCs. *See* Chlorofluorocarbons
Chain-of-custody (COC), 156–157
Charette, 16
Chartered Institution of Building Services Engineers (CIBSE), 188
 Application Manual 10:2005, Natural Ventilation in Non-Domestic Buildings, 188
Chlorofluorocarbons (CFCs), 124
CIBSE. *See* Chartered Institution of Building Services Engineers
CIRs. *See* Credit interpretation requests
Civil engineer, 13, 15
CMP. *See* Credential Maintenance Program
CO. *See* Carbon monoxide
CO_2. *See* Carbon dioxide
COC. *See* Chain-of-custody
Commingled waste. *See* Waste: Commingled

Commissioning (Cx), 120, 123
  enhanced, 110t, 121, 122t, 127, 129–130
  fundamental, building energy systems, 110t, 120–121, 122t, 125, 129–130
Commissioning authority (CxA), 13, 120–121
Community connectivity, 39t, 55, 65–66, 92–93
Compendium of Methods for the Determination of Air Pollutants, 192
Composite wood. *See* Wood: Composite
Construction:
  documents (CDs), 14–15, 121, 148
  IAQ management plan, 186t, 192–193, 207–208, 215–218
  processes, 9
  Specification Institute (CSI), 150–151
  waste management, 149t, 161–163, 170–171
Contaminants, 190, 192–193, 198. *See also* 4-phenylcyclohexene; Carbon dioxide (CO_2): emissions; Environmental tobacco smoke control; Formaldehyde; Particulates; Volatile organic compounds
Contractor, 13
Cost of Green Revisited by Davis Langdon, 257
Cradle-to-cradle (C2C) certified products, 244
Cradle-to-grave cycle, 152, 162
Credentialing system:
  application process, 5–7
  LEED AP with Specialty, 3–4
    LEED AP
      Building Design + Construction (BD+C), 4
      Homes, 4
      Interior Design + Construction (ID+C), 4
      Neighborhood Development (ND), 4
      Operations + Maintenance (O+M), 4

LEED Fellow, 3–5
LEED Green Associate, 3–4
reasons for earning credentials, 7
Credential Maintenance Program (CMP), 261
Credit interpretation requests (CIRs), 27, 30, 32–35
CRI. *See* Carpet and Rug Institute
CSI. *See* Construction: Specification Institute
Cx. *See* Commissioning
CxA. *See* Commissioning authority

D
Daylight and views:
  daylight, 185, 186t, 204–205, 239–240
  views, 186t, 205–206, 241–242
Daylighting, 185, 186t, 202–206, 239–242
Declarant, 30
Densely occupied spaces, 188–189
Development density, 39t, 55, 65–66, 92–93

E
EA. *See* Energy and Atmosphere
Emissivity, 47, 58
End users/occupants, 13, 185
Energy:
  cost budget, 118
  embodied, 152
  modeling, 16, 118
  regulated, 125–126
Energy and Atmosphere (EA), 22, 109–110
  categories, summary of, 268–269
  credits, 270t–271t
  energy consumption, tracking, 110, 120
  commissioning
    enhanced, EA credit 2, 110t, 121, 122t, 127, 129–130
    fundamental, building energy systems, EA prerequisite 1, 110t, 120–121, 122t, 125, 129–130

LEED compliance, strategies for, 120–122
measurement and verification, EA credit 3, 110t, 122–124, 128, 143–144
energy efficiency, 110–115
  energy performance, minimum, EA prerequisite 2, 110t, 115, 131–132
  energy performance, optimize
    equipment and appliances, EA credit 1.4, 110t, 118, 141–142
    HVAC, EA credit 1.3, 110t, 117–119, 139–140
    lighting controls, EA credit 1.2, 110t, 116, 119, 137–138
    lighting power, EA credit 1.1, 110t, 115–116, 118, 127, 135–136
  LEED compliance, strategies for, 115–118
prerequisites, 23, 270t–271t
refrigerants, management, 110, 124
  fundamental, EA prerequisite 3, 110t, 124–125, 127, 133–134
  LEED compliance, strategies for, 124
renewable energy
  Green Power, EA credit 4, 110t, 125, 127, 145–146, 248
Energy consumption, tracking, 110, 120–122
Energy efficiency:
  energy performance, minimum, 110t, 115, 131–132
  energy performance, optimize
    equipment and appliances, 110t, 118, 141–142
    HVAC, 110t, 117–119, 139–140
    lighting controls, 110t, 116, 119, 137–138
    lighting power, 110t, 115–116, 118, 127, 135–136
Energy Policy Act:
  of 1992 (EPAct 1992), 102, 104t, 115
  of 2005 (EPAct 2005), 102, 115
ENERGY STAR, 115, 118–119
Environmental:

performance, other quantifiable, 53–55, 90–91
Technology Verification (ETV), 196
tobacco smoke (ETS) control, 186t, 190, 198–199, 202, 211–212 (*See also* Contaminants)
EPA. *See* U.S. Environmental Protection Agency
EPAct:
    1992 (*See* Energy Policy Act of 1992)
    2005 (*See* Energy Policy Act of 2005)
EQ. *See* Indoor Environmental Quality
Equipment and appliances, 110t, 118, 141–142. *See also* ENERGY STAR
ETS. *See* Environmental tobacco smoke control
ETV. *See* Environmental Technology Verification
Exam, preparing for, 3–7, 21
    application process, 5–7
    prep guide
        BAIT tips, xii
            cheat sheet, xiii–xiv, 257–258
            flashcard tips, xii, xiv
            practice exam, xiv, 257–258
            structure, xi–xiv
    scoring of exam, 260
    structure of exam, 259–260
    study schedule, xiii
    study tips, xi–xii
    testing center environment, 278–279
    tips for taking, 260

F

Facility manager, 13
FAR. *See* Floor-to-area ratio
Fixtures:
    flow, 102, 104, 106
    flush, 102, 104
FloorScore Program, 194
Floor-to-area ratio (FAR), 40
Flushout, 193, 197
Fly ash, 157, 157t, 244
Footcandle, 48, 204, 206

Forest Stewardship Council (FSC), 153, 156, 159
Formaldehyde, 192. *See also* Contaminants
    phenol-formaldehyde, 194
    urea-formaldehyde, 194
FSC. *See* Forest Stewardship Council
FSC-certified wood products, 156, 157t
FTE. *See* Full-time equivalent
Full-time equivalent (FTE), 28
    occupancy, 61–62, 67, 102
Furniture and furnishings, 149t, 151, 174–175
    low-emitting materials, use in green building and, 186t, 194–197

G

GBCI. *See* Green Building Certification Institute
GC. *See* General contractor
General contractor (GC), 13, 15
GHG. *See* Greenhouse gas emissions
Glare control, 204–205
Glazing, 205
Global warming potential (GWP), 124
Good Practice Guide 237 by Carbon Trust, 188
Graywater, 51
Green building, 9. *See also* Sustainability
    benefits, 10–13 (*See also* Triple Bottom Line)
        economic, 11t, 11–12, 185, 202
        environmental, 11t, 11–12
        health and community, 11t, 11–12, 202
    costs, 16–17
Green Building and LEED Core Concepts Guide, 199
Green Building Certification Institute (GBCI), xi, 4–7, 17–18, 26–28, 30–32, 35
    website, 257–258, 261
Greenfield, 40, 55
GREENGUARD, 190, 196

Greenhouse gas (GHG) emissions, 11, 60, 159
Green Power, 110t, 125, 127, 145–146, 248
Green roofs, xii, 12, 43
Green Seal, 190
    Standard
        for Commercial Adhesives, GS-36, 194
        GC-3, 194
        GS-11, 194
Group multioccupant spaces, 204
Guidance on Innovation & Design (ID) Credits, 257
Guidelines for CIR Customers, 257
GWP. *See* Global warming potential

H

Heating, ventilation, air conditioning and refrigeration (HVAC&R), 121
Heating, ventilation and air conditioning (HVAC) systems, 124, 187, 198, 200
    strategies for energy performance, optimize, 110t, 113, 117–119, 139–140
Heat island effect, 44–45, 58–59
    nonroof, 45–46, 66, 76–77
    roof, 47–48, 65–66, 78–79, 249
HVAC. *See* Heating, ventilation and air conditioning systems
HVAC&R. *See* Heating, ventilation, air conditioning and refrigeration

I

IAQ. *See* Indoor air quality
ID+C. *See* LEED AP Interior Design + Construction
ID+C Reference Guide. See LEED Reference Guide for Green Interior Design and Construction
Impervious surfaces, 42–43
Incineration facilities, 159
Indoor air quality (IAQ), 185, 187
    minimum performance, 186t, 187, 209–210

Indoor chemical and pollutant source control, 186t, 198–199, 229–230
Indoor Environmental Quality (EQ), 12–13, 22
  categories, summary of, 269
  credits, 185, 189, 208, 249, 270t–271t
  indoor air quality, 185, 187
    construction management plan, 207
      during construction, EQ credit 3.1, 186t, 192, 215–216
      before occupancy, EQ credit 3.2, 186t, 192–193, 208, 217–218
    environmental tobacco smoke control, EQ prerequisite 2, 186t, 198–199, 202, 211–212
    indoor chemical and pollutant source control, EQ credit 5, 186t, 198–199, 229–230
    LEED compliance, strategies for, 187–188, 190–198
    low-emitting materials
      adhesives and sealants, EQ credit 4.1, 186t, 194, 196, 219–220
      composite wood and agrifiber products, EQ credit 4.4, 186t, 194, 225–226, 294
      flooring systems, EQ credit 4.3, 186t, 194, 223–224
      paints and coatings, EQ credit 4.2, 186t, 194, 221–222
      systems furniture and seating, EQ credit 4.5, 186t, 194–197, 227–228
    minimum performance, EQ prerequisite 1, 186t, 187, 209–210
    outdoor air delivery monitoring, EQ credit 1, 186t, 188–189, 213–214
    ventilation, increased, EQ credit 2, 186t, 188–189, 207–210
  lighting, 185, 202–204
    daylight and views
      daylight, EQ credit 8.1, 185, 186t, 204–205, 239–240
      views for seated spaces, EQ credit 8.2, 186t, 205–206, 241–242
    LEED compliance, strategies for, 204–205
    systems controllability, EQ credit 6.1, 186t, 204, 231–232
  prerequisites, 23, 189, 270t–271t
  thermal comfort, 185, 199, 201, 207
    design, EQ credit 7.1, 186t, 200–201, 235–236
    LEED compliance, strategies for, 199–201
    systems controllability, EQ credit 6.2, 186t, 199–200, 202, 233–234
    verification, EQ credit 7.3, 186t, 201–202, 237–238
Innovation in Design (ID), 22, 243–246
  exemplary performance, ID credit 1, 24, 53, 243–246, 250–251
  Energy and Atmosphere options, 115–116, 118, 125, 247, 274t
  Indoor Environmental Quality options, 205, 247–248, 275t
  Materials and Resources options, 151, 155–157, 244, 274t
  Sustainable Sites options, 53–54, 61, 243, 273t–274t
  Water Efficiency options, 151, 244, 274t
  LEED Accredited Professional on project team, ID credit 2, 243, 243t, 245, 247, 252–253
Integrated Project Delivery: A Guide, 16
Integrative project delivery (IPD), 13–16, 18–19
  benefits of design approach, 16, 110
  phases
    agency coordination/final buyout, 14–16
    certificate of occupancy, 14–15
    conceptualization, 14–15
    construction, 14–16
    criteria design, 14, 16
    detailed design, 14, 16
    final completion, 14–15
    implementation documents, 14, 16
    substantial completion, 14–15
International Organization for Standardization (ISO), 291
International Performance Measure & Verification Protocol (IPMVP), Volume III, 122, 124
IPD. *See* Integrative project delivery
IPMVP. *See* International Performance Measure & Verification Protocol, Volume III
ISO. *See* International Organization for Standardization
ISO 14021-1999: Environmental Label and Declarations, 154

L

Landfills, 159
Landscape architect, 13, 15
Langdon, Davis, 17, 257
LCA. *See* Life-cycle assessment
Leadership in Energy and Environmental Design (LEED). *See* LEED (Leadership in Energy and Environmental Design)
LEED (Leadership in Energy and Environmental Design), 3
  Accredited Professional (LEED AP), 243, 243t, 245
  AP (*See* LEED: Accredited Professional)
  AP Interior Design + Construction (ID+C), 3–4
    exam (*See* Exam, preparing for)
  certification levels, 24, 26, 247
  certification process, 26–33, 266t–267t
    appeals, 32–33, 35, 266t–267t
    certification review, 30–32, 31t, 34, 148, 267t
    construction review, 31, 266t
    credit form, 29–30, 34
    credit interpretation requests (CIRs) (*See* Credit interpretation requests)

design review, 30–31, 33, 266t
minimum program requirements (MPRs) (*See* Minimum program requirements)
project registration, 27, 33–34, 266t
split review, 31
CI (*See* LEED for Commercial Interiors)
consultant, 15
for Core & Shell, 25, 41
for Hospitality, 21
-Online, 26–32, 257–258
project administrator, 26
for Retail Interiors, 21
LEED AP Interior Design + Construction Candidate Handbook, 5, 17, 257, 259
LEED for Commercial Interiors (LEED CI), 21–22
 rating system, 39, 101, 109, 147, 185, 243, 257
  categories
   Energy and Atmosphere (EA) (*See* Energy and Atmosphere)
   Indoor Environmental Quality (EQ) (*See* Indoor Environmental Quality)
   Innovation in Design (ID) (*See* Innovation in Design)
   Materials and Resources (MR) (*See* Materials and Resources)
   Regional Priority (RP) (*See* Regional Priority)
   Sustainable Sites (SS) (*See* Sustainable Sites)
   Water Efficiency (WE) (*See* Water Efficiency)
  credit(s), 22–24, 270t–271t
   weightings, 24–26, 25t
  overview of, 21–25, 263t
  prerequisites, 22–23, 248, 270t–271t
  scorecard, 272
LEED for Interior Design and Construction Reference Guide. See LEED Reference Guide for Green Interior Design and Construction

LEED Reference Guide for Green Interior Design and Construction, 21–22, 104, 149, 154, 187, 194, 257
Life-cycle assessment (LCA), 17–18, 103, 115, 152, 202
Lighting, 185, 202–204
 controls, 110t, 116, 119, 137–138
 power, 110t, 115–116, 118, 127, 135–136
  density (LPD), 113, 115–116, 119
  reduction, 126, 128
 systems controllability, 186t, 204, 231–232
Light pollution reduction, 48, 58, 64, 80–81
Light trespass, 48
Low-emitting materials, 193
 adhesives and sealants, 186t, 194, 196, 219–220
 composite wood and agrifiber products, 186t, 194, 225–226, 294
 flooring systems, 186t, 194, 223–224
 paints and coatings, 186t, 194, 221–222
 systems furniture and seating, 186t, 194–197, 227–228
LPD. *See* Lighting: power: density
Luminare, 115

M

Maintenance, 9, 11, 24, 121
Mass transit, 61–62
Materials and Resources (MR), 22, 147–148
 categories, summary of, 269
 compliance for MR credits, 153, 158–159, 162, 248
 credits, 270t–271t
 material selection, 147, 152–153
  certified wood, MR credit 7, 149t, 153, 156–158, 163, 182–183
  LEED compliance, strategies for, 154–157
  rapidly renewable materials, MR credit 6, 149t, 153, 156, 158, 180–181

  recycled content, MR credit 4, 149t, 153–154, 157, 176–177
  regional materials, MR credit 5, 149t, 151, 153–155, 159, 178–179, 246
 prerequisites, 23, 270t–271t
 salvaged materials and material reuse, 147, 149, 206
  building reuse, interior nonstructural elements maintenance, MR credit 1.2, 149t, 150, 152, 168–169
  LEED compliance, strategies for, 150–151
  materials reuse
   furniture and furnishings, MR credit 3.2, 149t, 151, 174–175
   MR credit 3.1, 149t, 150–153, 172–173
 waste management, 147, 159–160
  construction waste management, MR credit 2, 149t, 161–163, 170–171
  LEED compliance, strategies for, 160–162
  recyclables, storage and collection of, MR prerequisite 1, 149t, 160, 163–165
  tenant space, long-term commitment, MR credit 1.1, 149t, 160–161, 166–167
Material selection, 147, 152–153
Materials reuse, 149t, 150–153, 172–173, 206
 furniture and furnishings, 149t, 151, 174–175
MDF. *See* Medium-density fiberboard
Measure and verification (M&V) plan, 122
Measurement and verification, 110t, 122–124, 128, 143–144
Mechanical, electrical and plumbing (MEP) systems, 13, 152
Medium-density fiberboard (MDF), 194
MEP. *See* Mechanical, electrical and plumbing systems
MEP engineer, 13–14
MERV filters, 190, 196, 198
Methane, 159, 162

Minimum efficiency reporting value (MERV) filters. *See* MERV filters
Minimum program requirement (MPR), 27–29, 32–33, 264t–265t
Montreal Protocol, 124
MPR. *See* Minimum program requirement
MPR 1:
   environmental law compliance, 28, 264t
MPR 2:
   permanency of building or space, 28, 264t
MPR 3:
   reasonable site boundary, 28, 264t
MPR 4:
   minimum floor area requirements compliance, 28, 265t
MPR 5:
   minimum occupancy rates compliance, 28, 265t
MPR 6:
   whole-building energy and water usage data sharing commitment, 28, 265t
MPR 7:
   minimum building area to site area ratio compliance, 28–29, 265t
MR. *See* Materials and Resources
M&V. *See* Measure and verification plan

N

National Institute of Standards and Technology (NIST), 24
Native plantings, 48–51
NIST. *See* National Institute of Standards and Technology
Nonpotable water. *See* Water: Nonpotable

O

Occupant(s), 13, 185
   satisfaction, 202, 207
Occupied spaces, 208
ODP. *See* Ozone-depleting potential
OPR. *See* Owner's project requirements
Oriented-strand board (OSB), 194
OSB. *See* Oriented-strand board
Outdoor air delivery monitoring, 186t, 188–189, 213–214
Owner, 13, 185
Owner's project requirements (OPR), 120–121, 123
Ozone-depleting potential (ODP), 124

P

Parking capacity, 39t, 62–64, 98–99
Particulates, 190, 192. *See also* Contaminants
Passive design strategies, 111
Pervious paving, 43
Pervious surfaces, 42–43
Phosphorous, total (TP), 43
Potable water. *See* Water: Potable
Public transportation access, 39t, 61, 63, 66, 94–95

R

Rapidly renewable materials, 149t, 153, 156, 157t, 158, 180–181
RECs. *See* Renewable energy certificates
Recycled content, 149t, 153–154, 157, 157t, 176–177
   assembly, 153
      postconsumer, 153–154, 157t
      preconsumer, 153–154, 157, 157t
Recycling, 149t, 160, 163–165
Reed, William, 257
Referenced standards, 23, 198, 248, 276t–279t
Refrigerants, management, 110, 110t, 124–125, 127, 133–134
Refurbished materials, 151
Regional materials, 149t, 151, 153–155, 157t, 159, 178–179, 246
Regional Priority (RP), 22, 245–247, 246t
Regional Priority credits (RPCs), 246
Remediation, 41. *See also* Brownfield: Redevelopment
   in-situ, 58
Renewable energy, 53, 110, 125
   Green Power (*See* Green Power)
   on-site, 53, 64, 88–89
Renewable energy certificates (RECs), 53, 125
Requests for information (RFIs), 15, 30
Retention ponds, wet or dry, 43
RFIs. *See* Requests for information
Roof design, 112

S

Salvaged materials, 147, 149–151
SCAQMD. *See* South Coast Air Quality Management District
Sheet Metal and Air Conditioning Contractor National Association (SMACNA), 190
Sick building syndrome, 187, 197
Site:
   conditions, 110
   location, 39
   selection, 9, 39, 39t, 41, 68–69
Skylights, 204–205
SMACNA. *See* Sheet Metal and Air Conditioning Contractor National Association
Solar heat gain, 204
Solar reflectance index (SRI), 44–46, 66, 112
South Coast Air Quality Management District (SCAQMD), 190
   SCAQMD Rule 1113, 194
   SCAQMD Rule 1168, 194, 197
SRI. *See* Solar reflectance index
SS. *See* Sustainable Sites
Stormwater:
   collection, 51–53, 60
   design
      quality control, 43–44, 65, 74–75
      quantity control, 43, 72–73
   management, 42–44
   runoff, 42, 65
Suspended solids, total (TSS), 43
Sustainability, 9. *See also* Green building

INDEX 303

Sustainable Building Technical Manual: Part II, 257
Sustainable design, xi–xii, 9, 56
Sustainable Sites (SS), 22, 39–40
　categories, summary of, 268
　credits, 270t–271t
　prerequisites, 270t–271t
　site selection, 40–41
　　brownfield redevelopment, SS credit 1, path 1, 41, 70–71
　　development density and community connectivity, SS credit 2, 39t, 55, 65–66, 92–93
　　environmental performance, other quantifiable, SS credit 1, path 12, 53–55, 90–91
　　heat island effect, 44–45, 58–59
　　　nonroof, SS credit 1, path 4, 45–46, 66, 76–77
　　　roof, SS credit 1, path 5, 47–48, 65–66, 78–79, 249
　　LEED compliance, strategies for, 41–55
　　light pollution reduction, SS credit, path 6, 48, 58, 64, 80–81
　　renewable energy, on-site, SS credit, path 11, 53, 64, 88–89
　　SS credit 1, 39t, 41, 68–69
　　stormwater design
　　　quality control, SS credit 1, path 3, 43–44, 65, 74–75
　　　quantity control, SS credit 1, path 2, 43, 72–73
　　wastewater technologies, innovative, SS credit 1, path 9, 51–53, 84–85
　　water efficient landscaping, SS credit 1, paths 7 and 8, 50–51, 58–60, 82–83
　　water use reduction, SS credit, path 10, 53, 86–87
　transportation, 60
　　bicycle storage and changing rooms, SS credit 3.2, 39t, 61–62, 62t, 64–65, 67, 96–97
　　LEED compliance, strategies for, 60–62
　　parking capacity, SS credit 3.3, 39t, 62–64, 98–99

　　public transportation access, SS credit 3.1, 39t, 61, 63, 66, 94–95
Systems performance testing, 121

T

TAGs. *See* Technical advisory groups
Technical advisory groups (TAGs), 18
Tenant space, 15, 39–41, 65. *See also under* LEED for Commercial Interiors rating system categories
　long-term commitment, 149t, 160–161, 166–167
　occupancy, 102, 105
Thermal comfort, 185, 199, 201, 207
　design, 186t, 200–201, 235–236
　systems controllability, 186t, 199–200, 202, 233–234
　verification, 186t, 201–202, 237–238
Thermal emittance, 45, 59
Thermal energy storage, 111
Tobacco smoke. *See* Environmental tobacco smoke control
Tools for the Reduction and Assessment of Chemical and Other Environmental Impacts (TRACI), 24
Total:
　phosphorous (TP), 43
　suspended solids (TSS), 43
TP. *See* Total phosphorous
TRACI. *See* Tools for the Reduction and Assessment of Chemical and Other Environmental Impacts
Traditional project delivery, 13–15
　phases
　　agency permit/bidding, 14–15
　　certificate of occupancy, 14–15
　　construction, 14–15
　　construction documents (CDs), 14–15
　　design development, 14–15
　　final completion, 14–15
　　predesign/programming, 14
　　schematic design, 14
　　substantial completion, 14–15

Transient users, 61
Transmitted light (T_{vis}), 204
Transportation:
　factors impacted by
　　fuel, 60, 63
　　human behavior, 60, 63
　　location, 60
　　vehicle technology, 60, 63
　public (*See* Public transportation access)
Transportation Management Plan, Comprehensive, 61
Triple Bottom Line, 11t, 11–12
　Indoor Environmental Quality, impacts on, 187, 202
　Sustainable Sites, impacts on, 39
TSS. *See* Total suspended solids
T_{vis}. *See* Transmitted light

U

U.S Energy Information Administration, 60
U.S. Environmental Protection Agency (EPA), 147, 185
USGBC. *See* U.S. Green Building Council
U.S. Geological Survey, 101
U.S. Green Building Council (USGBC), 5–6, 17–18, 24, 27–28
　Regional Councils, 245
　website, 3, 10–11, 30, 109, 245, 257–258

V

Value engineering (VE), 16
VE. *See* Value engineering
Vegetated swales, 43
Ventilation, 186t, 187–189, 207–210
　mixed-mode, 187–188
　rates, 187
VOCs. *See* Volatile organic compounds
Volatile organic compounds (VOCs), 190, 192–193, 197. *See also* Contaminants
VOC budget method, 193, 193t

W

Waste:
  commingled, 160
  management, 10, 147, 159–160
Wastewater technologies, innovative, 51–53, 84–85
Water:
  consumption, 24, 104t
  heating or cooling, 111
  nonpotable, 108
  potable, 48–49, 51, 101, 106
Water Efficiency (WE), 22, 24, 101–102
  categories, summary of, 268
  credits, 270t–271t
  indoor water use
    water use reduction, WE credit 1, 102t, 105–108
    water use reduction, WE prerequisite 1, 102t, 104–108
  prerequisites, 23, 270t–271t
Water efficient landscaping, 48–51, 58–60, 82–83
Waterless urinals, 103
Water use reduction, 53, 86–87, 102t, 102–108
WE. *See* Water Efficiency
Wetlands, 42
Wood:
  certified, 149t, 153, 156–158, 163, 182–183
  composite, 186t, 194, 225–226, 294

X

Xeriscaping, 48, 51

Z

Zoning, 29

Sample Flashcards

1

Q. What is the percentage threshold of the tenant space used to determine compliance with EA Credit 3: Measurement and Verification?

2

Q. What are the CSI MasterFormat™ divisions to include in the MR credit calculations to prove compliance?

3

Q. What is sustainable design?

4

Q. What savings have green buildings achieved?

5

Q. What are the seven strategies to improve IAQ during construction?

6

Q. What are the two strategies to improve IAQ by prevention and segregation methods?

2

A. CSI MasterFormat Divisions 03–10, plus Foundations and Sections, Paving, Site Improvements, and Plantings. Mechanical, electrical, and plumbing equipment or any specialty items, such as elevators, are not included in any of the calculations. Division 12, Furniture must be included as well.

1

A. 75 percent of total building area

4

A.
- Up to 50 percent energy use reduction
- 40 percent water use reduction
- 70 percent solid waste reduction
- 13 percent reduction in maintenance costs

3

A. According to the Brundtland Commission of the United Nations' website: development that meets the needs of the present without compromising the ability of future generations to meet their own needs.

6

A.
- EQ Prerequisite 1: Environmental Tobacco Smoke (ETS) Control
- EQ Credit 5: Indoor Chemical and Pollutant Source Control

5

A.
- EQ Credit 3.1: Construction IAQ Management—During Construction
- EQ Credit 3.2: Construction IAQ Management—Before Occupancy
- EQ Credit 4.1: Low-Emitting Materials—Adhesives and Sealants
- EQ Credit 4.2: Low-Emitting Materials—Paints and Coatings
- EQ Credit 4.3: Low-Emitting Materials—Flooring Systems
- EQ Credit 4.4: Low-Emitting Materials—Composite Wood and Agrifiber Products
- EQ Credit 4.5: Low-Emitting Materials—Systems Furniture and Seating

7

Q. What are the three ways to earn Innovation in Design Credits?

8

Q. What are the benefits of an integrated project delivery (IPD)?

9

Q. What are the 3 requirements to comply with EA Credit 3: Measurement and Verification should the tenant more than 75 percent of the total building area?

10

Q. What four components are required for a team to submit for an Innovation in Design credit?

11

Q. What are the two compliance path options for EA Credit 3: Measurement and Verification should a tenant lease less than 75 percent of the total building area?

12

Q. What are the three compliance path options for EQ Credit 4.5: Low-Emitting Materials, Systems Furniture and Seating?

13

Q. What are the four compliance path options for EQ Credit 8.1: Daylight and Views, Daylight?

14

Q. What is the minimum percentage of regularly occupied spaces that must receive daylight in order to comply with EQ Credit 8.1: Daylight and Views, Daylight?

8

A.
- Holistic approach versus linear approach with conventional projects.
- Brings all of the project team members together early to work collectively.
- Begins as early as possible in the design process.
- Team members share the risk and the rewards.
- Helps to lower first costs and operating costs.
- BIM.
- Multilateral agreements.

7

A.
- Exemplary Performance
- Innovation in Design
- LEED Accredited Professional

10

A.
- Intent of strategy
- Suggested requirements for compliance
- Suggested documentation proving compliance with requirements
- A narrative describing the strategy implemented

9

A.
- Install continuous metering equipment.
- Develop and implement a measure and verification (M&V) plan in accordance with the International Performance Measure & Verification Protocol (IPMVP), Volume III.
- Provide an approach for corrective action should the anticipated energy savings not be achieved during operation.

12

A.
- GREENGUARD Indoor Air Quality Certified products
- Furniture and seating that has passed the EPA Environmental Technology Verification (ETV) Large Chamber Test Protocol for Measuring Emissions of VOCs and Aldehydes Test
- Furniture and seating to pass the ANSI/BIFMA M7.1 2007 and ANSI/BIFMA x7.1-2007 testing protocols

11

A.
- Install submetering equipment for all energy sources (2 points).
- Energy costs should be paid by the tenant (3 points).

14

A. 75 percent for one point and 90 percent for 2 points

13

A.
- Simulation
- Prescriptive
- Measurement
- Combination

15

Q. What is the minimum percentage of regularly occupied spaces that must have a view to the exterior environment in order to comply with EQ Credit 8.2: Daylight and Views, Views for Seated Spaces?

16

Q. What are the four strategies required to comply with EQ Credit 5: Indoor Chemical and Pollutant Source Control?

17

Q. What is the requirement difference between the two compliance options for MR Credit 5: Regional Materials?

18

Q. What are the two differences between prerequisites and credits?

19

Q. What two components are credit weightings based on?

20

Q. What are the certification levels and coordinating point ranges for LEED?

21

Q. What is the percentage threshold of the tenant space to determine compliance with SS Credit 3.3: Alternative Transportation, Parking Availability?

22

Q. Describe the MPR for whole-building energy and water use data.

16

A.
- 10-foot long entryway system
- Exhaust high chemical areas
- MERV 13 filters or better
- Containment of hazardous liquid waste

15

A. 90 percent

18

A.
- Prerequisites are mandatory, as they address minimum performance achievements, and credits are optional.
- Prerequisites are not worth any points, whereas credits are.

17

A. Option 1 only requires products to be *manufactured* within 500 miles from the project site, whereas Option 2 requires the materials to also be *extracted* and *processed* within 500 miles.

20

A.
- Certified: 40–49 points
- Silver: 50–59 points
- Gold: 60–79 points
- Platinum: 80 and higher

19

A. Environmental impacts and human benefits

22

A. All certified projects must commit to sharing with USGBC and/or GBCI all available actual whole-project energy and water usage data for a period of at least five years starting from the date that the LEED project begins typical physical occupancy. LEED CI projects that do not have metering for the tenant space is exempt from compliance.

21

A. 75 percent of total building area

23

Q. What are the minimum SRI values for steep-sloped and low-sloped roofs?

24

Q. What are the six required tasks to comply with EA Prerequisite 1: Fundamental Commissioning?

25

Q. What is the project size limitation to be aware of when pursuing EA Credit 2: Enhanced Commissioning if an independent third-party will not be hired?

26

Q. What is considered a previously developed site?

27

Q. What are the additional five tasks to be completed to be in compliance with EA Credit 2: Enhanced Commissioning?

28

Q. How long can temporary irrigation systems be used for?

29

Q. What must be achieved in order for a project to be eligible to pursue LEED certification?

30

Q. What is the default value to use when calculating electricity consumption?

24

A.
1. Select a CxA
2. Document OPR and BOD.
3. Include Cx requirements in CDs
4. Commissioning Plan
5. Verify installation of energy-related systems
6. Summary Cx report

23

A.
- Low-sloped (≤ 2:12) = 78
- Steep-sloped (> 2:12) = 29

26

A. Properties with existing buildings and hardscape but also those that were graded or somehow altered by human activity

25

A. 50,000 square feet

28

A. 1 year maximum

27

A.
1. Review at mid-construction documents
2. Review submittals
3. Systems manual
4. Training
5. Revisit project within 8 to 10 months after substantial completion

30

A. 8 kw/sf/yr

29

A. Complying with all minimum program requirements, achieving all prerequisites, and earning a minimum of 40 points

31

Q. What is the default recycled content value of steel?

32

Q. What is the time frame in which trees can mature in order to provide shade?

33

Q. In order to comply with EA Credit 2: Enhanced Commissioning when must the CxA return to the tenant space after substantial completion for a review with the O&M staff?

34

Q. What is an LPE?

35

Q. What are the six characteristics of credit interpretation requests?

36

Q. Which comes first, the OPR or BOD?

37

Q. What are the two factors to address within the SS category?

38

Q. What is floor-to-area ratio?

32 **A.** 5 years	**31** **A.** 25 percent postconsumer
34 **A.** Licensed-professional exemption—the path decided on a Credit form to reduce documentation requirements	**33** **A.** Within 8 to 10 months
36 **A.** The owners project requirements come before the basis of design is generated.	**35** **A.** <ul><li>Issued after a project is registered</li><li>Issued for a fee</li><li>Credit interpretation ruling issued in response</li><li>Submitted for clarification referencing *one* credit or prerequisite</li><li>Ruling not final</li><li>Submitted via LEED-Online</li></ul>
38 **A.** The proportion of the total floor area of a building to the total land area the building can occupy	**37** **A.** <ul><li>Site Selection</li><li>Transportation</li></ul>

39

Q. What are the two strategies of Site Selection?

40

Q. What is community connectivity?

41

Q. What is development density?

42

Q. What is a brownfield?

43

Q. What are the two factors to be addressed to be in compliance with SS Credit: Site Selection, Option 2, Path 3, Stormwater Design, Quality Control?

44

Q. Which two credits require a MERV filter? What filter is required for each?

45

Q. How much does the outdoor air supply need to be increased beyond EQ Prerequisite 1: Minimum Indoor Air Quality Performance to comply with EQ Credit 2: Increased Ventilation? What is the referenced standard?

46

Q. If a project cannot comply with ASHRAE 62.1-2007, how might the tenant still comply with EQ Prerequisite 1: Minimum Indoor Air Quality Performance and still seek certification?

40

A. Proximity of project site to local businesses and community services such as parks, grocery stores, banks, cleaners, pharmacies, and restaurants. The LEED rating systems require a connection to at least 10 basic services.

39

A.
- SS Credit 1: Site Selection
- SS Credit 2: Development Density and Community Connectivity

42

A. Real property, the expansion, redevelopment, or reuse of which may be complicated by the presence or potential presence of a hazardous substance, pollutant, or contaminant. Cleaning up and reinvesting in these properties protects the environment, reduces blight, and takes development pressures off green spaces and working lands.

41

A. The total square footage of all buildings within a particular area, measured in square feet per acre or units per acre

44

A.
- EQ Credit 3.1: Construction Indoor Air Quality Management Plan – During Construction
- EQ Credit 5: Indoor Chemical and Pollutant Source Control

43

A. 80 percent Total suspended solids (TSS) and 40 percent Total phosphorus (TP)

46

A. The team would need to document why the tenant space cannot comply and then prove an absolute minimum of 10 cubic feet per minute per person is supplied.

45

A. Exceed ASHRAE 62.1-2007 by at least 30 percent

47

Q. What is the referenced standard for brownfield sites?

48

Q. What are the four impacts of transportation?

49

Q. When should a survey be conducted to comply with EQ Credit 7.2: Thermal Comfort – Verification?

50

Q. What are the three SS credits to reduce the transportation impacts associated with the built environment?

51

Q. What other credit must a tenant pursue should they also wish to pursue EQ Credit 7.2: Thermal Comfort – Verification?

52

Q. In which two means is stormwater management addressed?

53

Q. What are native and adaptive plantings?

54

Q. What is potable water?

48

A.
- Location
- Vehicle technology
- Fuel
- Human behavior

47

A. ASTM E1903-97

50

A.
- SS Credit 3.1: Public Transportation Access
- SS Credit 3.2: Bicycle Storage and Changing Rooms
- SS Credit 3.3: Parking Availability

49

A. Within 6 to 18 months after the space is occupied

52

A. Quantity Control and Quality Control

51

A. EQ Credit 7.1: Thermal Comfort – Design

54

A. Drinking water supplied by municipalities or wells

53

A. Native vegetation occurs naturally, whereas adaptive plantings are not natural; they can adapt to their new surroundings. Both can survive with little to no human interaction or resources.

55

Q. What is imperviousness?

56

Q. What is perviousness?

57

Q. What is stormwater runoff?

58

Q. What is a footcandle?

59

Q. What is the heat island effect?

60

Q. What is emissivity?

61

Q. What is solar reflective index (SRI) and the associated scale?

62

Q. What is albedo and the associated scale?

56

A. Surfaces that allow at least 50 percent of water to percolate or penetrate through them

55

A. Surfaces that do not allow 50 percent or less of water to pass through them

58

A. A measurement of light measured in lumens per square foot

57

A. Rainwater that leaves a project site flowing along parking lots and roadways, traveling to sewer systems and water bodies

60

A. The ratio of the radiation emitted by a surface to the radiation emitted by a black body at the same temperature

59

A. Heat absorption by low-SRI, hardscape materials that contribute to an overall increase in temperature by radiating heat

62

A. The ability to reflect sunlight based on visible, infrared, and ultraviolet wavelengths on a scale from 0 to 1

61

A. A material's ability to reflect or reject solar heat gain measured on a scale from 0 (dark, most absorptive) to 100 (light, most reflective).

63

Q. What is a building footprint?

64

Q. What are the three credits that address salvaged materials and materials reuse?

65

Q. What are the three strategies to manage stormwater?

66

Q. What are the seven steps of energy modeling?

67

Q. What are the referenced standards used for indoor water use?

68

Q. What is baseline versus design case?

69

Q. What are examples of flow fixtures?

70

Q. How are flow fixtures measured?

64

A.
- MR Credit 1.2 Building Reuse, Maintain Interior Nonstructural Elements
- MR Credit 3.1: Materials Reuse
- MR Credit 3.2: Materials Reuse, Furniture and Furnishings

63

A. The amount of land the building structure occupies, not including landscape and hardscape surfaces such as parking lots, driveways, and walkways

66

A.
1. Select a modeler.
2. Determine the portion of the building to model (as it will typically include more than the tenant space).
3. Select a modeling approach (depending if the base building is already modeled).
4. Acquire building information such as as-built drawings, operational schedules, HVAC system and zones, and lighting systems.
5. Model the design case.
6. Model the baseline case using the referenced standards mandatory provisions and prescriptive requirements.
   a. Model an alternative baseline case (only if the base building is more efficient than the reference standard.
7. Calculate the energy reduction.

65

A.
- Minimize impervious areas
- Control stormwater
- Harvest rainwater

68

A. The amount of water a conventional project would use as compared to the design case

67

A. EPAct 1992 and EPAct 2005

70

A. Gallons per minute (gpm)

69

A. Sink faucets, showerheads, and aerators

71
Q. What are examples of flush fixtures?

72
Q. How are flow fixtures measured?

73
Q. What is graywater?

74
Q. What are the LEED ID+C strategies to address reducing indoor water consumption within the WE category?

75
Q. What are the strategies to reduce outdoor water consumption?

76
Q. How is compliance calculated for each MR credit? Area or cost?

77
Q. To comply with EA Credit 1.2: Optimize Energy Performance – Lighting Controls what percentage of the connected lighting power must have daylight responsive controls?

78
Q. How many compliance paths are available under SS Credit 1: Site Selection, option 2 should the tenant space not be located in a LEED certified base building?

72

A. Gallons per flush (gpf)

71

A. Toilets and urinals

74

A.
- WE Prerequisite 1: Water Use Reduction
- WE Credit 1: Water Use Reduction

73

A. Wastewater from showers, bathtubs, lavatories, and washing machines. This water has not come into contact with toilet waste, according to the International Plumbing Code (IPC).

76

A.
- MR Credit 1.2: Building Reuse = Area
- MR Credit 3.1: Material Reuse, MR Credit 4: Recycled Content, MR Credit 5: Regional Materials, MR Credit 6: Rapidly Renewable Materials = Cost
- MR Credit 3.2: Material Reuse, Furniture = Furniture Cost
- MR Credit 7: Certified Wood = Wood Cost

75

A.
- Implement native and adapted plants
- Use xeriscaping
- Specify high-efficiency irrigation systems
- Use nonpotable water for irrigation

78

A. 12

77

A. 50 percent

79

Q. What are the benefits of utilizing a CxA?

80

Q. What is the baseline standard for energy performance?

81

Q. What are the uses for regulated energy?

82

Q. What are the uses for process energy?

83

Q. What two components should be evaluated when determining which refrigerants to use?

84

Q. What is ODP?

85

Q. What is GWP?

86

Q. What is light pollution?

80

A. ASHRAE 90.1–2007, Appendix G

79

A.
- Minimize or eliminate design flaws.
- Avoid construction defects.
- Avoid equipment malfunctions.
- Ensure that preventative maintenance is implemented during operations.

82

A. Computers, office equipment, kitchen refrigeration and cooking, washing and drying machines, and elevators and escalators. Miscellaneous items, such as a waterfall pumps, and lighting that is exempt from lighting power allowance calculations (for example, lighting integrated into equipment) are also categorized as process energy uses.

81

A.
- Lighting: interior and exterior applications (parking garages, facades, site lighting)
- HVAC: space heating, cooling, fans, pumps, toilet exhaust, ventilation for parking garages
- Domestic and space heating for service water

84

A. Ozone-depleting potential

83

A. Refrigerants should be evaluated based on ODP and GWP impacts.

86

A. Waste light from building sites that produces glare, is directed upward to the sky, or is directed off the site

85

A. Global warming potential

87

Q. What are the four components of the Energy & Atmosphere category?

88

Q. What is the strategy that addresses the management of refrigerants within the EA category?

89

Q. To comply with EA Credit 1.2: Optimize Energy Performance – Lighting Controls what percentage of the connected lighting power must have occupancy sensors?

90

Q. What are the three strategies that address the tracking of energy consumption?

91

Q. What are the six types of eligible renewable energy sources for LEED projects?

92

Q. What is the opportunity in which a tenant can incorporate renewable energy into a project seeking LEED CI certification?

93

Q. What is the most sustainable strategy to reduce waste?

94

Q. What are the five opportunities to address energy performance for a LEED project?

88

A. EA Prerequisite 3: Fundamental Refrigerant Management

87

A.
- Energy efficiency
- Tracking energy consumption
- Managing refrigerants
- Renewable energy

90

A.
- EA Prerequisite 1: Fundamental Commissioning
- EA Credit 2: Enhanced Commissioning
- EA Credit 3: Measurement and Verification

89

A. 75 percent

92

A. Purchase green power, or RECs (EA Credit 4: Green Power).

91

A. Solar, wind, wave, biomass, geothermal power, and low-impact hydropower

94

A.
- EA Prerequisite 2: Minimum Energy Performance
- EA Credit 1.1: Optimize Energy Performance – Lighting Power
- EA Credit 1.2: Optimize Energy Performance – Lighting Controls
- EA Credit 1.3: Optimize Energy Performance – HVAC
- EA Credit 1.4: Optimize Energy Performance – Equipment & Appliances

93

A. Source reduction. An example includes purchasing products with less packaging, as there is less waste to dispose of.

95

Q. What are the three components to address within the Materials & Resources category?

96

Q. What is a rapidly renewable material?

97

Q. Define recycled content.

98

Q. What referenced standard declares a material having postconsumer/preconsumer recycled content?

99

Q. Recycled material that was generated from a manufacturing process is referred to as _____. What are some examples?

100

Q. Recycled material that was generated by household, commercial, industrial, or institutional end users, which can no longer be used for its intended purpose is referred to as _____.

101

Q. What are considered regional materials according to LEED?

102

Q. What type of documentation is required to prove compliance for FSC wood?

96 **A.** Fiber or animal materials that must be grown or raised in 10 years or less	**95** **A.** ■ Salvaged materials and material reuse ■ Building material selection ■ Waste management
98 **A.** ISO 14021-1999—Environmental Label and Declarations	**97** **A.** The percentage of materials in a product that are recycled from the manufacturing waste stream (preconsumer waste) or the consumer waste stream (postconsumer waste) and used to make new materials. Recycled content is typically expressed as a percentage of the total material volume or weight.
100 **A.** Postconsumer. Examples include construction and demolition debris, materials collected through recycling programs, and landscaping waste (ISO 14021).	**99** **A.** Preconsumer. Examples include planer shavings, sawdust, bagasse, walnut shells, culls, trimmed materials, over issue publications, and obsolete inventories. Not included are rework, regrind, or scrap materials capable of being reclaimed within the same process that generated them.
102 **A.** FSC wood requires chain-of-custody documentation.	**101** **A.** The amount of a building's materials that are extracted, processed, and manufactured close to a project site, expressed as a percentage of the total material cost. LEED considers regional materials as those that originate within 500 miles of the project site.

103

Q. Which elements are excluded from the calculations for recycled content, regional materials, and rapidly renewable materials?

104

Q. What are the four credits that address strategies to reduce the impacts of building material selection?

105

Q. What are the minimum types of items to be recycled during operations to meet the requirements of the MR prerequisite?

106

Q. What are the 3R's of waste management in order of hierarchy?

107

Q. What are the three strategies to reduce waste?

108

Q. What are the three components discussed in the EQ category?

109

Q. What are the types of VOCs?

110

Q. What are the different referenced standards for products that emit VOCs?

104

A.
- MR Credit 4: Recycled Content
- MR Credit 5: Regional Materials
- MR Credit 6: Rapidly Renewable Materials
- MR Credit 7: Certified Wood

103

A. Mechanical, electrical, and plumbing equipment, and hazardous waste materials

106

A. Reduce, reuse, and recycle

105

A. Paper, corrugated cardboard, glass, plastics, and metals

108

A.
- Indoor air quality (IAQ)
- Thermal comfort
- Lighting

107

A.
- MR Prerequisite 1: Storage & Collection of Recyclables
- MR Credit 1.1: Tenant Space, Long-Term Commitment
- MR Credit 2: Construction Waste Management

110

A.
- Paints—Green Seal 11, Green Seal 3
- Coatings—SMACNA 1113
- Adhesives—Green Seal 36
- Sealants—SCAQMD 1168
- Carpet—Carpet and Rug Institute (CRI) Green Label Plus Program
- Other Flooring—FloorScore

109

A. Volatile organic compounds include carbon dioxide, tobacco smoke, and particulates emitted from carpet, paints, adhesives, glues, sealants, coatings, furniture, and composite wood products.

111
Q. How are VOCs measured?

112
Q. To comply with EQ Credit 4.4: Low-Emitting Materials—Composite Wood and Agrifiber Products, these products are not allowed to contain what type of resin? What type of resin is acceptable?

113
Q. What is MERV? What is the range?

114
Q. What are the three ventilation strategies to address IAQ?

115
Q. What are the four environmental factors of thermal comfort defined by ASHRAE 55?

116
Q. What are the three credits that address thermal comfort?

117
Q. What are the EQ credit strategies that address lighting within a LEED certified tenant space?

118
Q. How long is the minimum lease term to comply with MR Credit 1.1: Tenant Space, Long-term Commitment?

112

A. Urea-formaldehyde resin is not allowed, although phenol-formaldehyde resin is.

111

A. Grams per liter (g/L)

114

A.
- EQ Prerequisite 1: Minimum Indoor Air Quality Performance
- EQ Credit 1: Outdoor Air Delivery Monitoring
- EQ Credit 2: Increased Ventilation

113

A. Minimum Efficiency Reporting Value (MERV) filters range from 1 (low) to 16 (highest).

116

A.
- EQ Credit 6.2: Controllability of Systems, Thermal Comfort
- EQ Credit 7.1: Thermal Comfort, Design
- EQ Credit 7.2: Thermal Comfort, Verification

115

A. Humidity, air speed, air temperature, and radiant temperature

118

A. 10 years

117

A.
- EQ Credit 6.1: Controllability of Systems, Lighting
- EQ Credit 8.1: Daylight and Views, Daylight
- EQ Credit 8.2: Daylight and Views, Views for Seated Spaces

119

Q. Of the six available Regional Priority credits, how many can count toward a project's LEED certification?

120

Q. A project's available Regional Priority points are determined by what project-specific element?

121

Q. In order to find out the Regional Priority points for a LEED project, what must be done first?

122

Q. How many Innovation in Design points can a LEED Green Associate earn for a project seeking LEED certification?

123

Q. A project team diverts 97 percent of the nonhazardous construction waste away from a landfill. They earn all the points that are within the credit and an additional point that will be included in which category?

124

Q. Where must CO_2 sensors be placed?

125

Q. At what level of CO_2 concentration is typically harmful?

126

Q. If a tenant agrees to pay for their utility usage, where might they apply this strategy within the LEED CI rating system?

127

Q. What are the five SMACNA guidelines?

128

Q. A 42,225 square foot tenant space is interested in purchasing RECs to comply with EA Credit 4: Green Power. If they are seeking exemplary performance, how much should they purchase?

| **120** | **119** |
| A. Zip code | A. Four |

| **122** | **121** |
| A. Zero, only LEED APs are eligible. | A. Refer to the USGBC website and find your project's region. |

| **124** | **123** |
| A. Between 3 and 6 feet above the floor | A. Innovation in Design |

| **126** | **125** |
| A. EA Credit 3: Measurement and Verification, should they lease less than 75 percent of the total building area. | A. 530 parts per million (ppm) |

| **128** | **127** |
| A. 675,600 kWh | A. HVAC protection, source control, pathway interruption, housekeeping, and scheduling |